COAL

The Earth series traces the historical significance and cultural history of natural phenomena. Written by experts who are passionate about their subject, titles in the series bring together science, art, literature, mythology, religion and popular culture, exploring and explaining the planet we inhabit in new and exciting ways.

Series editor: Daniel Allen

In the same series

Coal

Ralph Crane

REAKTION BOOKS

For Joy, Callum and Rhiannon

Published by Reaktion Books Ltd
Unit 32, Waterside
44–48 Wharf Road
London N1 7UX, UK
www.reaktionbooks.co.uk

First published 2021

Printed and bound in India by Replika Press Pvt. Ltd

A catalogue record for this book is available from the British Library

ISBN 978 1 78914 366 9

CONTENTS

Preface

In the opening chapter of his curious little book *The History of a Lump of Coal: From the Pit's Mouth to a Bonnet Ribbon*, published in 1882, Alexander Watt observes that 'of all the substances which are dug out of the earth, coal is one of the most remarkable in the vast extent of its usefulness to man'.[1] The subtitle, which signals the breadth of Watt's topic, also anticipates the cultural history of the present volume.

As we shall see in the following pages, coal and humankind share a history that stretches back millennia. It is likely that the Awabakal, the Aboriginal people from the Lake Macquarie region of New South Wales, burnt coal many centuries ago. The Reverend L. E. Threlkeld records that the name for Lake Macquarie in the local language is '*Nik-kin-ba*, from *Nikkin*, Coal, and *ba* place of, meaning a place of coal'.[2] The Awabakal are also believed to be the only Aboriginal people to refer to coal in their Dreamings. And whether we are cognizant of it or not, coal continues to play a big role in our lives today, both literally and metaphorically. Coal was used as a source of heat by ancient cave dwellers, it fuelled the Industrial Revolution, and today it is the number-one energy resource used to generate electricity globally, as well as a major contributor to climate change. Indeed, it is hard to overestimate the impact of coal on our modern world.

While the physical presence of coal in our lives is not as apparent as it was even half a century ago, when much of the Western world still relied on coal fires for home heating, the metaphorical presence still lingers from the age of coal and

Chinese poster, 1973. The text (in both calligraphy and modern style) reads 'I have at my disposal precious mineral deposits hidden in the far mountains.'

before. The word 'coal' is used in a figurative sense in the Bible on several occasions (though the references are, of course, likely to have referred to the embers of wood, rather than mineral coal). The quenching of the coal of a man in 2 Samuel 14:7 refers to the annihilation of his children, while the meaning of heaping coals of fire on the head of an enemy in Proverbs 25:22 and Romans 12:20 is (somewhat counterintuitively) a kindness bestowed on an enemy. In the Middle Ages coal was associated with disease, death and the Devil: the buboes of the Black Death were described as resembling 'broken fragments of brittle sea-coal'; the word 'carbuncle' is derived from the Latin term for a live coal; and 'anthrax' is taken from the Greek name for charcoal.[3] As Barbara Freese notes ironically, 'in the Middle Ages coal had quite an image problem' – as it does today, albeit for very different reasons.[4] We continue to use the metaphorical vocabulary of coal in everyday idioms such as 'at the coalface' (to be directly engaged in something), 'a canary in a coal mine' (to provide advance warning of danger), 'haul over the coals' (to scold or reprimand severely) and 'carrying coal to Newcastle' (performing a useless activity). And the capitals of England and Scotland owe their nicknames – the 'Big Smoke' and 'Auld Reekie' (Scots for Old Smokey) – to the air pollution caused by the burning of coal that frequently left a choking smog hanging overhead. Indeed, the narrator of Jules Verne's *The Child of the Cavern* (1877) maintains that the Scottish capital's 'nickname of "Auld Reeky" is justified by its smoke-laden atmosphere'.[5]

Given its importance in the Industrial Revolution, it is of little surprise that coal featured prominently at the Great Exhibition of 1851. The Britain and the Empire exhibit, for example, included a block of parrot (or cannel) coal from the West Wemyss Colliery in Fifeshire and a garden seat made out of the same material (both exhibited by HRH Prince Albert in Class 27, Manufactures from Mineral Substances),[6] while the Van Diemen's Land (Tasmania) and New Zealand areas also included coal specimens. Similarly, the Belgian exhibit included examples of several varieties of coal from the Belgian coalfield. And so on.

Auguste Charles Pugin, 'Coal Exchange', from Thomas Rowlandson and A. C. Pugin, *The Microcosm of London; or, London in Miniature* (1808–10, plate 17).

Today, coal continues to attract attention, though largely of a more negative ilk. The Adani Group's proposed Carmichael Coal Mine in central Queensland, Australia, for instance, has generated significant controversy surrounding its financial viability, its claimed economic benefits and, most importantly, the impact it would have on the environment, including on the iconic Great Barrier Reef.

This book discusses the nature and culture of coal. It addresses the meaning of coal over time, explores some of the myriad ways coal has shaped and continues to shape the history of humankind, and discusses the significant influence it has had on literature and the arts. It tells *a* story of coal, rather than *the* story of coal: as John Berger so eloquently puts it in his novel *G.* (1972), 'Never again will a single story be told as though it were the only one.'

Entrance to the Anthracite Heritage Museum, Pennsylvania.

1 What Is Coal?

We all recognize it, but what exactly is the solid black stuff that in Jules Verne's *The Mysterious Island* (1874) Gideon Spillet labels 'the most precious of minerals'?[1]

Coal is the official state mineral of Kentucky; it is also the official state rock of both Utah and West Virginia. So is it a mineral or is it a rock? The answer is not entirely straightforward, as a cursory survey of dependable dictionaries illustrates. According to the *Oxford English Dictionary* coal is 'a hard, opaque, combustible black or blackish mineral' (Kentucky); the *Collins English Dictionary* defines it as 'a combustible compact black or dark-brown carbonaceous rock' (Utah and West Virginia); and the *Merriam-Webster Dictionary* hedges its bets, describing coal as 'a black or brownish black solid combustible substance' (Utah, West Virginia *and* Kentucky). Mineral; rock; substance.

To further compound the issue, in legal terms coal is defined as a mineral in the United States ('the term "mineral" includes all inorganic substances, as well as hydrocarbons, such as oil and natural gas, and carboniferous deposits, such as coal'[2]). However, while coal has both vegetable and mineral attributes, in the geological lexicon (in the United States as elsewhere) it is classed as a type of rock. Rock appears to beat mineral in the definition game; Utah and West Virginia pip Kentucky. In a 1956 article which specifically addresses the lack of a clear, consistent and technically valuable definition of coal, James M. Schopf defines coal as:

> a readily combustible rock containing more than 50 percent by weight and more than 70 percent by

volume of carbonaceous material, formed from compaction or induration of variously altered plant remains similar to those of peaty deposits. Differences in the kinds of plant materials (type), in degree of metamorphism (rank), and range of impurity (grade), are characteristic of the varieties of coal.[3]

Cigarette cards: from set of fifty 'Mining' cards issued by Wills's Cigarettes in 1916.

Likewise, coal has been defined as 'a combustible rock that had its origin in the accumulation and physical and chemical alteration of vegetation'.[4] More recently, Geoscience Australia defines coal in similar terms: 'a combustible sedimentary rock formed from ancient vegetation which has been consolidated between other rock strata and transformed by the combined effects of microbial action, pressure and heat over a considerable time period'.[5] Coal differs from other kinds of rock, however, in that it is composed of organic carbon, the actual remains of dead plants.

But coal is never just coal. As H. Stanley Jevons remarks, 'to the dealer in coal it is only the name of a whole class of substances, and . . . means about as much as the words "cloth" or "paper"'.[6] Coal is identified by type, which is a measure of organic composition; by grade, which is a measure of its mineral matter content; and, perhaps most importantly for the layperson, by rank, which is a measure of the degree of coalification.

Because coal is made up of organic matter rather than minerals, geologists and petrologists use the analogous term 'macerals', proposed by the British botanist (and women's rights campaigner) Marie C. Stopes and adopted in 1935, to refer to the components of coal.[7] There are three major groups of macerals. Liptinites (previously called exinites) are dark grey in colour. They are derived from spores, pollens and resins in the original plant material. Vitrinites are medium to light grey in colour, and the most abundant of the three groups. They are formed from cell walls and woody tissues such as bark and roots. Inertinites are white and sometimes very bright. They are largely oxidation products of other macerals, and richer in carbon than either liptinites or vitrinites.[8]

Coal seam, Guido Mine, Zabrze, Poland.

Based on its maceral content and macroscopic appearance, coal can be classified into four principal rock types, which 'reflects the nature of the plant debris from which the original organic matter was derived'[9]: vitrain, clarain, durain and fusain. Vitrain, composed largely of vitrinite, has a bright lustre and tends to be brittle. Clarain has a less brilliant appearance than that of vitrain, and is made up of alternating bright layers of vitrinite and duller layers of liptinite and inertinite. Durain has a hard granular texture and is usually a dull black or dark grey colour. Fusain closely resembles charcoal. It is soft and crumbly and made up of inertinites.[10]

While coal can be as old as 2 billion years or as young as 2 million years, the formation of the vast majority of the world's coal occurred during the Carboniferous (or coal-bearing) period, a geologic period spanning 60 million years from roughly 359 to 299 million years ago. During this period, vast, dense swamps or wetlands covered much of Earth's tropical land, and as the plant matter from these forests died, it sank to the floor of the swamp where it decayed and formed peat. The coalification process – the geological process through which vegetal matter is transformed first into peat and then gradually into coal, largely as a result of temperature and pressure – extends over millions of years. As the Australian novelist Kate Grenville poetically puts it, 'Eventually . . . those squillions of dead trees turned into the hard shiny black rock called coal.'[11]

The degree of change undergone during the coalification process determines the physical and chemical properties of coal and hence the rank of coal, ranging from peat to anthracite, which is a measure of the level of transformation undergone during the formation process from vegetal matter to carbon. As Jennifer M. K. O'Keefe et al. succinctly explain, 'Coal rank is generally considered to be a function of some combination of heat, pressure, and time.'[12] In most countries, including the

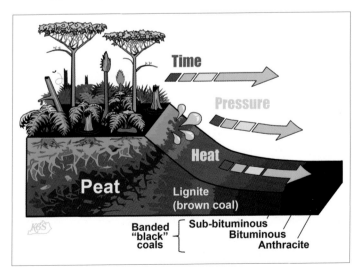

Coal rank and how coal forms (Kentucky Geological Survey).

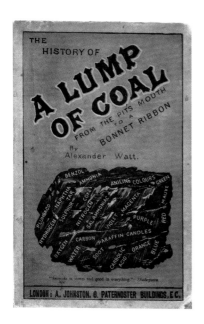

Cover design for Alexander Watt's *A Lump of Coal: From the Pit's Mouth to a Bonnet Ribbon* (London, 1882).

United Kingdom, the four principal ranks of coal are peat, lignite, bituminous and anthracite. In the classification system used in the United States and Canada, peat is considered a precursor to coal and the four ranks of coal are lignite, sub-bituminous, bituminous and anthracite. There are also subdivisions within these ranks – thermal and metallurgical coal are subdivisions of bituminous coal, for example. It is worth noting, however, that these classification systems employ arbitrary distinctions, and that, like the classifications used by coal dealers, the divisions used in different countries are a matter of convenience rather than real or objective ones. Generally, the deeper in the ground the coal is found, the higher its rank (Hilt's law), and the higher the rank the higher the carbon content and the lower the water content. Peat is closest to the surface and has the highest moisture content, while anthracite is the hardest grade of coal and tends to be found deeper underground than other grades. Lignite is known as brown coal, while sub-bituminous coal, bituminous coal and anthracite are collectively referred to as black coal.

As the first step in the geological formation of coal, peat is a very soft brown coal – often considered a pre-coal rather than a true coal – composed of over 60 per cent organic material, and because of its high water content its uses as a fuel are limited.

Lignite, or brown coal, is friable, has a high moisture content, and typically contains between 25 and 35 per cent carbon. Alexander Watt suggests it 'has the appearance of being only partially mineralized or converted into coal'. 'Lignites', he continues, 'are distinguished from "true coals" by their giving but little flame, while yielding abundance of smoke.'[13] In geological terms it is young coal, formed somewhere between 250 million years ago and the present. It ranges in colour from brown to black, and has a low heating value or coal energy rating. Lignite beds are found relatively near the surface, are often thick, and are comparatively easy to mine. However, despite the lower costs of mining, its high moisture content and low energy value make it uneconomical to transport, and lignite power stations – such as the Belchatów power station in Poland, the world's largest lignite-fired power station, or the Loy Yang B power station in Australia's Latrobe Valley – tend to be built close to the mines that supply them with brown coal.

Sub-bituminous coal (sometimes called black lignite) is slightly older, harder and blacker in colour than lignite. It also has a higher carbon content, typically between 35 and 45 per cent, and a higher energy rating than lignite. However, its energy rating is well below that of bituminous coal, which means that more coal needs to be burned to produce the same amount of energy as bituminous coal, and its high sulphur content makes it more damaging to the environment than harder coals.

Bituminous coal is formed as more pressure is applied to lignite/sub-bituminous seams, expelling more water and increasing the carbon content to between 45 and 85 per cent. It is both harder and blacker than lignite and sub-bituminous coal, and appears shinier. This rank of coal, which includes steam (or thermal) coal and coking (or metallurgical) coal, is the most abundant form of coal and has a higher heating value (Btu) than lignite or sub-bituminous coal, making it ideal for use in the steel and iron industries, as well as in power plants generating electricity.

Anthracite, also commonly referred to as hard coal, accounts for just 1 per cent of the world's coal reserves, and is mined in only a few countries. It is the oldest, hardest and blackest type of coal, and unlike all other coal types, which are classified as organic sedimentary rocks, it is often classified as a metamorphic rock. It has a carbon concentration of over 85 per cent, and a slightly higher energy rating than bituminous coal. Clean to the touch, anthracite's high heat production makes it a particularly valuable type of coal. It has been mined in South Wales since the medieval period, and was first found in the United States in 1790 in what would become known as the 'Coal Region' of eastern Pennsylvania. Today China mines about 75

Small coal mine near
Matarrosa del Sil,
León Province,
Spain, 1984.

per cent of the global output of anthracite. Watt informs his readers that anthracite

> differs greatly from all other species of coal. It burns without flame or smoke, and yields an intense heat, leaving but little ash. It does not burn well in an ordinary fireplace, unless there is a good draught in the chimney. This coal is used largely in America, and is generally kindled with charcoal, fresh fuel being required not oftener than once or twice a day. A fire made with anthracite coal requires no poking. It is a very good and economical plan to mingle a little anthracite with ordinary coal for domestic purposes.[14]

Technically, coal is a fossil fuel which is continually being formed, and therefore a renewable resource. However, as the coalification process takes millions of years, it is generally classed as a non-renewable resource. Like other similar natural resources, including crude oil and natural gas, coal simply cannot be replaced as quickly as it is being used. Moreover, as coal resources are further depleted over the next several hundred years they may become too costly or too difficult to mine; however, long before the future supply of coal becomes an issue, consumption is likely to decline as a result of increasing concerns about the climate, and we may choose to shift our reliance to renewable resources such as solar power or wind power.

We have long been aware of the detrimental effects – to both people and the environment – of burning coal, whatever its rank. Indeed, activists in Britain have been warning of the problems associated with coal smoke since the early 1840s.[15] In an 1882 issue of *Popular Science Monthly*, C. William Siemens comments on the smoke caused by burning raw coal that envelops London and other large towns in Britain, and the concomitant threat to public health:

> Professor Roberts has calculated that the soot in the pall hanging over London on a winter's day amounts to fifty tons, and that the carbonic acid [carbon monoxide], a poisonous compound, resulting from the imperfect combustion of coal, may be taken as at least five times that amount. Mr Aitken has shown, moreover . . . that the fine dust resulting from the imperfect combustion of coal is mainly instrumental in the formation of fog . . . The hurtful influence of smoke upon public health, the great personal discomfort to which it gives rise, and the vast expense it indirectly causes through the destruction of our monuments, pictures, furniture, and

John Hassell, *Coal Works: A View near Neath in Glamorganshire in South Wales*, 1798, aquatint.

Coal barge passing through the Grand-Carré lock on the Deûle canal, Lille, France, 2014.

apparel, are now being recognized, as evidenced by the success of recent Smoke Abatement Exhibitions.[16]

Public meetings, pamphlets and exhibitions all contributed to public awareness of 'the destructive effects of the smoky and inefficient domestic fireplace'.[17] An early Smoke Abatement Exhibition, organized in 1881–2 by what would become the National Smoke Abatement Institution, showcased both domestic and industrial smoke abatement appliances. The exhibition attracted 116,000 people in South Kensington (including the

Detail from the mural in homage to the lignite mining industry in Cerceda, Spain.

Prince of Wales), and a further 32,000 people when it shifted to Manchester. Similar exhibitions continued to be held regularly in towns across the country until at least the 1950s, with visitors being encouraged to switch from coal fires to smokeless technologies such as gas and electricity as those fledgling industries grew.[18] Yet in 1938 coal still provided heat in 80 per cent of all households in Britain, with families essentially living and eating in a single room heated by an open coal fire.[19] In 'The Case for the Open Fire', a 1945 essay published in the *Evening Standard*, George Orwell, while recognizing the contribution of coal fires to the fogs in British cities, nevertheless makes a powerful case for the pivotal position of the coal fire in the family home. He lauds the 'aesthetic appeal' of the coal fire, and argues that 'an open fire makes for sociability', and 'that every house or flat should have at least one open fire round which the family can sit'.[20] However, following the Great Smog of December 1952, when for five days London was smothered in a deadly fog that killed thousands, arguments in favour of domestic coal use became increasingly difficult to sustain. Further, in 1938 the amateur meteorologist Guy Callendar – building on earlier work by Irish physicist John Tyndall (in 1861) and Nobel Prize-winning Swedish scientist Svante Arrhenius (in 1896) – had published a

Coal seam, State Coal Mine, Wonthaggi, Victoria, Australia.

paper in the *Quarterly Journal of the Royal Meteorological Society* that demonstrated that the Earth's land temperature had increased over the last half-century. He also drew a direct link between the burning of coal and other fossil fuels and what would become known as global warming – though he did not appreciate the negative impacts of this climate change.[21]

Notwithstanding our knowledge of its damaging effects on the environment, and the shift away from coal in the domestic hearth, coal remains a vital source of industrial energy in many countries. Australia, and to a lesser extent the United States, for example, both remain reliant on coal for their electricity needs. Several European countries including Germany and Poland have recently built or plan to build new coal power stations. Similarly, both India and China continue to rely heavily on coal-fired power stations despite concerns about rising carbon dioxide levels and the Dickensian air pollution in major cities including Delhi and Beijing that is being caused by coal burning in surrounding regions.

Coal is part of the Earth. It has played a major role in human history, and has fuelled some of the greatest industrial achievement of humankind. Through much of human history coal has been viewed as a resource to be exploited. Over the last two hundred years it has enriched the few while enthralling the masses. It has been an agent of progress for local, national and global societies, and it has been an agent of harm to people and communities around the world. In *Fuel: A Speculative Dictionary*, Karen Pinkus succinctly sums up the nature of coal: 'Coal – burned in one-room huts or teepees by Native Americans, in the homes of the bourgeois and the miners in northern France, in the factories of London, in Dickens's fictional Coketown, even in home furnaces in parts of the world, to say nothing of industry – is, it goes without saying, dirty.'[22]

As we better understand 'coal's sweeping impact on the global environment and on society',[23] our challenge in the future is to learn to live in a world that can no longer be dependent on the black stuff.

2 Using Coal

In Peter Paul Rubens's painting *Old Woman with a Basket of Coal* (*c.* 1618–20), the titular female warms her hands above a basket of burning coal, a boy blows on the embers and a youth looks into the fire. The flames from the smouldering coal light the faces of the three figures, as well as providing them with some warmth. Rubens's painting can be read as an allegory of the dependence on coal for heat and light across the three ages – childhood, adulthood and old age. And, albeit more often indirectly, that dependence continues today.

Perhaps surprisingly to us in the twenty-first century, coal was valued for its aesthetic appearance long before it was used as a source of fuel. Six thousand years ago the inhabitants of northeastern China, like the Romans in Britain much later, valued coal for its looks rather than as something that could help protect them from the cold. Archaeological evidence has firmly established that Neolithic inhabitants of the Shenyang area of northeastern China were carving ornaments, including what appear to be ear-piercing ornaments, from black lignite (jet) as far back as 4000 BCE, and that 'By the Warring States period [475–221 BCE], this handicraft industry was flourishing over large areas of China . . . and it continued to do so through the Han and later.'[1]

While it is impossible to know exactly when humankind first used coal, China was certainly in the vanguard of coal use in the ancient world, not only for ornament but for fuel. What is believed to be the earliest working coal mine in the world, the Fushun mine in northeastern China, probably began operating

Mural in the
Anthracite
Heritage Museum,
Pennsylvania.

around 3000 BCE, well ahead of similar developments in other parts of the world. And it is quite conceivable that coal was first used as a fuel in China at a very early date, given the likelihood of the Chinese seeing coal outcrops burning, or wood fires accidentally setting alight their combustible lignite carving materials. However, the first hard evidence we have for coal being used as a fuel dates from the Han dynasty (206 BCE to 220 CE). By the final years of the Han dynasty coal was definitely being used for both domestic and industrial purposes, including heating and the production of iron, though it is difficult to ascertain how widely it was used. By the time of the short-lived Sui dynasty (581–618 CE) there are written accounts that confirm coal was being used to cook meat and warm wine, though again, there is no evidence to indicate how widely it was being used for these purposes.[2]

By the late eleventh century, when a shortage of timber led to rapid growth in coal production and use, 'coal had become the exclusive fuel used in well over 100,000 households in the capital, Khai-feng', and it was widely used in the silkworm industry to maintain temperatures in the huts and to heat the water needed for the production process.[3] Two hundred years later, in his celebrated account of his travels in Asia, the Venetian explorer Marco Polo provides this wonderfully quirky description of coal and its uses in the Chinese region of Cathay:

Peter Paul Rubens,
*Old Woman with a
Basket of Coal*, 1618–20,
oil on panel.

Throughout this province there is found a sort of black stone, which they dig out of the mountains, where it runs in veins. When lighted, it burns like charcoal, and retains the fire much better than wood; insomuch that it may be preserved during the night, and in the morning be found still burning. These stones do not flame, excepting a little when first lighted, but during their ignition give out a considerable heat. It is true there is no scarcity of wood in the country, but the multitude of inhabitants is so immense, and their stoves and baths, which they are continually heating, so numerous, that the quantity could not supply the demand; for there is no person who does not frequent the warm bath at least three times a week, and during the winter daily, if it is in their power. Every man of rank or wealth has one in his house for his own use; and the stock of wood must soon prove inadequate to such consumption; whereas these stones may be had in greatest abundance, and at a cheap rate.[4]

Polo's account records only the domestic applications of coal in northern China – its use in cooking stoves and for heating public or private baths. Nevertheless, by the time of his visit at the end of the thirteenth century, coal was also widely used for a range of industrial purposes, including firing ceramic kilns and iron smelting. In the following centuries the use of coal spread to southern China, and the industrial uses of coal, particularly as furnace fuel in the iron industry, became considerably more important than the household ones observed by Marco Polo.

While he was evidently familiar with charcoal, a fuel produced by burning carbon woods, it appears Polo was unfamiliar with the 'black stone', or mineral coal, he saw in China. But coal use was already well established in Britain by the time of his journey east. There is evidence that the Bronze Age people in Wales used coal on their funeral pyres, and that the Romans used coal as a fuel.[5] When the Romans invaded Britain between 40 and 50 CE it is likely that coal was included under the name of jet, which they carved and polished into jewellery.[6] However, as Freese explains, 'The Romans occupying Britain did more

John Ruskin, *View of a Colliery at the Edge of a Town*, 1840–49, watercolour over graphite on card.

with coal than merely dress up with it; they began burning it, too. Soldiers burned coal in their forts, blacksmiths burned coal in their furnaces, and priests honored Minerva, the goddess of wisdom, by burning coal in the perpetual fire at her shrine in Bath.[7] Yet while there is archaeological evidence of coal use by the Romans in Britain – in stations along Hadrian's Wall, in the Roman baths at Lanchester or the remains of a Roman smithy at Ebchester, for example – there is little evidence of widespread coal use by the Romans, and whatever uses they did make of coal seem to have come to an end when they left in the fifth century.[8] One of the few mentions of coal use during the Dark Ages, which followed Roman rule, can be found in the *Ecclesiastical History of England*, written around 731 CE by the Venerable Bede, who notes that 'Britain . . . has much and excellent jet, which is black and sparkling, glittering at the fire, and when heated, drives away serpents'.[9] Ironically, Bede wrote his chronicle at his monastery in Monkwearmouth, in Northumbria, in what would turn out to be one of the richest coal regions of the country. If his reference is to be taken at face value, in eighth-century Britain coal was burnt to provide smoke to ward off snakes, rather than to provide heat.

Moving forward through the years, there is evidence of coal being dug in the coalfield of Zwickau in Saxony in the tenth century, and some evidence of coal being used as a fuel in Britain during the Norman period. More widespread use of coal in Britain, and simultaneously in Belgium, however, seems to have begun towards the end of the twelfth century, though it would still be some time before coal was used widely in Britain and Europe and a coal industry as we know it could be said to exist.[10] Evidence of the use of coal outside the coalfields and a growing coal trade in the early thirteenth century includes the duty payable on coal unloaded at London's Billingsgate during Henry III's reign (1216–72); a street in London, near Ludgate Circus, which was given the name Seacole Lane sometime before 1228; and the death of Robert le Portour, who drowned while unloading coal in the Thames in 1236.[11]

In the thirteenth century coal was gathered or dug in many parts of Britain – around Colne in Lancashire, and Swansea in South Wales, for example – for use as a fuel, much of it by smiths. Indeed, J. U. Nef maintains in his epic work *The Rise of the British Coal Industry*, 'that during the Middle Ages there were only two artisans who made any considerable use of coal – the limeburner and the smith'.[12] As the fledgling industry grew and demand expanded beyond the coalfields, so the coal trade developed. Records suggest that sea coal (mineral coal as distinct from charcoal) gathered at Plessey, a village in Northumberland, was shipped from the River Blyth to London in the early decades of the century, and that by the middle of the century the coal trade began in earnest along what the poet John Milton calls the 'coaly Tyne',[13] from where it would be shipped to London and other markets of eastern England. (Five hundred years later a young James Cook learned his seamanship in the North Sea coal trade, sailing on collier cats and barks for nine years, before enlisting in the Royal Navy.[14] Cook's ship HMS *Endeavour*, used on his first voyage of discovery to the Pacific, 1768–71, was originally built as a collier.) While the coal trade at this time was growing on the Tyne, it would remain relatively insignificant for another two hundred years for the simple reason that coal use was trifling, in part because of the slow take-up of coal as a domestic fuel due to the smoke it created and its smell. The absence of chimneys in all but manor houses and castles meant that the coal burnt in domestic hearths would slowly fill the rooms with smoke. Outside the home complaints about air pollution caused by burning coal reached such a level that in 1285 a series of commissions were set up in London to address the problem, and by 1306 objections to the heavy use of coal in London by blacksmiths, brewers and the like led to the use of coal being banned altogether.[15] Nevertheless, coal use increased as the century progressed, the population grew and forests were depleted – though the Black Death, which killed half the population of Britain, caused a temporary slump.

Coal use increased in the fourteenth century, and the coal trade grew rapidly in the northern coalfields of Northumberland and Durham, with numerous collieries established to meet demand. Staithes (landing stages) were built along the River Tyne near Newcastle for the loading and also the storing of coal, which was transferred by keels (flat-bottomed boats) to coal-carrying sailing ships to be sent to other parts of the country or to Calais (the only port outside the country to which sea coal could be exported at this time). At the same time the number of 'landsale' collieries, which catered for a growing inland demand for coal for domestic use, also grew. The fourteenth century likewise saw the rise of coal production in other parts of the country, including the Lancashire, Yorkshire, Derbyshire, Nottinghamshire and Staffordshire coalfields. The increase in coal consumption continued through the fifteenth century as new uses arose: coal was employed in place of wood or peat in the manufacture of salt from sea water, for example, and in lime burning.[16] Coal was also used extensively in the (albeit small) medieval armament industry in Europe, from Scotland to Germany, France and Belgium.[17]

William Gray, writing in 1649, signals the next step towards the age of coal: 'Coales in former times was onely used by smiths, and for burning of lime; woods in the south parts of England decaying, and the city of London, and other cities and townes growing populous, made the trade for coale increase yearely.'[18] During the sixteenth and seventeenth centuries coal was transformed from a source of heat used only in particular localities and industries into the principal fuel for much of the nation. From the beginning of the sixteenth century, as open fireplaces in the centre of a room began to be replaced by fireplaces with flued chimneys that carried away the coal smoke, the demand for coal for domestic purposes slowly outstripped the artisanal demand from smiths, lime burners and fledgling attempts to

'A Coal Mine: Miners at Work above and below Ground', coloured lithograph.

smelt ores. Then in the 1570s domestic coal use in Britain surged – until by the end of the century it had become London's staple fuel. Within another couple of decades, as lingering objections to the smell of coal smoke dissipated, coal was widely burned in the homes of the rich as well as the poor, and by the end of the seventeenth century London's heavy coal use made it the most polluted of all European cities.

At this juncture London relied on coal shipped by sea from the coalfields of the northeast. The narrator of Michael Ondaatje's novel *Warlight* (2018) observes that 'Christopher Wren constructed St Paul's Cathedral but also converted the lower reaches of the Fleet, broadening its borders so it could be used for transporting coal.'[19] In fact Wren's proposal was rejected in favour of a plan to convert the Fleet into a canal, completed in 1680. Newcastle Close and Old Seacoal Lane (off Farringdon Street) are both named for the coal wharves, used in the coastal coal trade, that once lined the Fleet River. By 1700 coal consumption was significantly greater than it had been a century and a half earlier (up from 177,000 tons at the beginning of Queen Elizabeth I's rule to as much as 3 million tons a year), and Britain was mining significantly more coal than the rest of the world combined. In pre-Industrial Revolution Britain coal was still being used primarily to help 'the British protect themselves from the seasons', but other industrial uses had also been found for the vast reserves of energy in Britain's coalfields, including 'brick-making, glass, ceramics, soapboiling, lime burning, forging, distilling, and brewing'.[20] In fact coal appears to have been used successfully to manufacture glass in the Newcastle area from about 1619. And late in the century Martin Ele gave an account to the Royal Society of making pitch, tar and oil from pit coal in the Shropshire coalfield.[21] The textile industry, which used coal principally for heating the vats, cauldrons and kilns employed at various stages of the manufacturing process, was also one of the largest industrial users of coal in the seventeenth century. However, not everyone welcomed the rapidly increasing use of coal during this period. Queen Elizabeth I herself expressed her dislike of the taste and smoke of sea coal, and many visitors to

London in the seventeenth century (including the diarist John Evelyn, who published *Fumifugium; or, The Inconveniencie of the Aer and Smoak of London Dissipated* in 1661) complained about the smell of the unwholesome smoke that hung over the city.[22]

In the early decades of the eighteenth century coal continued to be used in the salt industry, although by the middle of the century that once thriving industry had passed its heyday. But then as the use of coal in one industry declined, other industries grew or came along which demanded even more coal. Undoubtedly, the two most significant developments in the use of coal in the 1700s were related to the metal industry and the advent of steam engines. The use of the coal-derivative coke in the iron industry, pioneered by Abraham Darby in 1709, was firmly established during the course of the century and rapidly took Britain from a country dependent on imported iron to the world's most efficient producer of iron, the supply of which would prove key to both its industrial expansion at home and the expansion of its empire abroad.[23] And steam engines – perhaps the most important invention ever in terms of both the production and the consumption of coal – 'would prove to be the pumping heart of the industrial revolution'.[24] Coal no longer only supplied heat, it also supplied motion; the heat produced by coal could now be converted into motive power. As Heidi C. M. Scott explains: 'Earth's most abundant fossil fuel, coal, powered the dynamos of the Industrial Revolution beginning in the eighteenth century. The union of coal with steam engines changed the world.'[25] By the beginning of the Industrial Revolution Britain already had well-advanced maritime, canal and road networks, and rudimentary railways with wagons drawn by horses – all initially developed or upgraded to facilitate the transportation of bulky cargoes of coal from the coalfields to the manufacturing centres.

Coal use began, of course, in the collieries themselves, where slack coal was used to feed the boilers of pumping and winding engines.[26] As the nineteenth century unfolded Britain became utterly dependent on coal. Instead of small water-powered and largely manually operated mills, industry shifted to large coal-powered and mechanized factories, and the lives of the workers who toiled in them were shaped both directly and indirectly by the use of coal. As George Orwell argues in *The Road to Wigan Pier* (1937), 'Our civilization, *pace* Chesterton, *is* founded on coal, more completely than one realizes until one stops to think about it. The machines that keep us alive, and the machines that make the machines, are all directly or indirectly dependent upon coal.'[27] (Orwell appears to be responding to G. K. Chesterton's proposal that 'civilisation is founded upon abstractions'.[28]) By the middle of the century coal was used to make the iron that was in turn used to build the machines that were used in factories throughout the country; coal was used to provide the power that ran the machines and factories; coal gas was used to provide light for the workers; coal was used to fire the bricks that were used to build both the factories and the houses to accommodate the labour forces; and at home the workers and their families cooked their food over coal fires, which also provided their only form of heating. Every aspect of the lives of factory workers in the burgeoning industrial towns and cities of Britain was shaped by the use of coal; at home and at work, the world of the working classes 'was constructed, animated, illuminated, colored, scented, flavoured, and generally saturated by coal and the fruits of its combustion'.[29] The monotonous mill towns of Lancashire – including Manchester (nicknamed 'Cottonopolis' in the nineteenth century, while correspondingly Wigan was known as 'Coalopolis'), Oldham, Bolton and Preston – were built as much on the coal that fuelled their factories as on the cotton they produced.

Lancashire offers a useful example to illustrate the ubiquity of coal. Indeed, it would be fair to say that directly and indirectly coal drove the economic development of Lancashire (as it did other parts of Britain) in myriad ways during the eighteenth and nineteenth centuries. First, of course, was the abundance of accessible coal measures in the Lancashire coalfield (sometimes referred to as the 'cotton coalfield') – the most productive outside the northeast until it was overtaken by South Wales in 1883. The burgeoning coal industry fed the factories of the mill towns, and led to the development of the canals, which were constructed to transport coal from the pits to the towns and ports of the

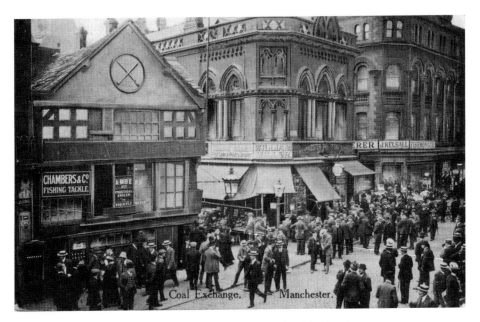

county. The Sankey Brook Navigation, the first canal in England, which opened in 1757, was built to smooth the supply of coal from the pits around Haydock to the industries they fed in Liverpool, while the first section of the Bridgewater Canal, which opened a few years later in 1761, was built to transport coal from the Duke of Bridgewater's pits in Worsley to Manchester. Coal fuelled Liverpool's salt boilers as well as domestic demand, and by the end of the eighteenth century coal exports to Ireland and America were among the staples stimulating the growth of the town. The emergence of Liverpool as a pioneer in the growth of seagoing steam navigation as well as the increased export trade with Ireland was fuelled by coal from the St Helens coalfield. Coal also stimulated maintenance of good road transport and the growth of the railways in Lancashire. The combination of coal resources, good transport systems and ready access to markets encouraged the growth of other industries on the back of coal. Ironworks were developed on a number of coalfields; copper smelting was important for a time; and glass, chemical (notably alkali) and soap industries flourished. By the 1860s the Lancashire coalfield was producing somewhere in the region of

Vintage postcard: Coal Exchange, Manchester, 1916.

12 million tons per year, a third of which was consumed by the steam engines which powered the cotton factories. By the end of the century there were around 350 collieries in the Lancashire coalfield, collectively producing nearly 20 million tons annually. Quite simply, the growth of manufacturing and commerce in Lancashire in the eighteenth and nineteenth centuries was made possible by coal. And vice versa. In the early twentieth century, as coal exports declined, the Lancashire coalfield was somewhat cushioned, albeit temporarily, by the fact that much of the coal it produced was consumed locally.[30]

The geography of coal was crucial to its use in the development of manufacturing industries and the growth of manufacturing centres. And while coalfields, and consequently manufacturing towns, were spread across the length and breadth of Britain, in the United States the geography of coal favoured certain places. Pittsburgh grew rapidly to become a major manufacturing centre, with glass and iron among its earliest industries, thanks to the enormous deposits of bituminous coal discovered under Mount Washington (earlier called 'Coal Hill'). As Freese notes, 'In Britain, it had taken centuries for areas to go from forested wilderness to industrial metropolis. Pittsburgh, like so much of what would become the United States, would experience that history in concentrated form, propelled in no small part by the concentrated energy beneath its hills.'[31] By 1832, almost all the factories in Pittsburgh ran on cheap coal-powered steam, and as coal use in Pittsburgh grew, so did its pollution levels, rapidly making it the most polluted city in the western hemisphere.[32] Meanwhile, in the early decades of the nineteenth century, the cities of the Eastern Seaboard had no easy access to coal or cheap steam power. That changed with the development of 'coal's liquid pathways', a series of river and canal systems that linked the rural anthracite-producing regions of Pennsylvania with the urban markets of

Poster promoting Pennsylvania, showing head and shoulders of a coal miner, by Isadore Posoff, 1937.

the Eastern Seaboard, enabling the development of an industrial economy based on coal that would have a massive impact on life in the mid-Atlantic states.[33] Over the next one hundred years coal fuelled the new industrial capitalism and the transformation of America from an economy based on small-scale enterprise to one dominated by giant corporations.[34]

The nineteenth century was witness, too, to other coal-using inventions. Gas lighting using coal gas was introduced into factories (enabling longer working days) and theatres, and later into homes and for street lighting. Pall Mall in London was lit by coal gas on 28 January 1807, and on 23 August 1821 St James's Park was lit by coal gas for the first time.[35] In 1816, Preston in Lancashire became the first British town outside London to be lit by coal gas (a blue plaque in Fox Street commemorates the first building to be lit by piped gas), and by the middle of the century most provincial towns in Britain with a population of more than 2,500 had gas lighting.[36] Gas lighting spread through Europe, with Paris getting street lighting in 1820. Baltimore was the first city in the United States to instal coal-gas street lighting, and in Australia Sydney's streets were lit by gas for the first time in 1841.

Mauveine, the first aniline or coal-tar dye, was discovered accidentally by William Perkin in 1856. Perkin's purple dye quickly became popular, and he established a factory in London, giving birth to the synthetic dye industry. Coal tar, an abundant by-product of making coal gas and coke, was also the main ingredient of the distinctive yellow-orange bars of Wright's Coal Tar Soap, developed by William Valentine Wright four years later, in 1860. Today the coal by-products of the original formula have been removed, and the soap has been rebranded as Wright's Traditional Soap with coal-tar fragrance – presumably to lend the new product a whiff of nostalgia.

As far back as the sixteenth century the Royal Navy drew on the country's merchant marine in times of war, including ships plying the coal trade – which was seen by many as the chief nursery for English seamen.[37] The long relationship between coal and national defence intensified on 12 July 1871 when HMS *Devastation* (the ship represented on the 'England's Glory' matchbox) was launched. HMS *Devastation* was the first ship in the Royal Navy to be propelled solely by coal-powered steam engines, and when working at full power it could consume up to 150 tons of coal in a day. As coal-powered steamships replaced sailing ships, the Navy became dependent on coaling stations in the principal ports of the British Isles and across the empire. The need to be able to protect these coaling stations signalled the importance of coal in the defence of the empire as Britain entered the period of high imperialism, which began in the last quarter of the century and lasted through to the end of the First World War.[38] A 1915 magazine advert for Wright's Coal Tar Soap declared that 'BRITAIN'S MIGHT IS (W)RIGHT'. The advertisement, an example of what Anne McClintock calls commodity racism,[39] shows a lion, its paw resting on a large block of Wright's Coal Tar Soap, in front of a Union Jack fluttering on what looks like a ship's jackstaff. The advert clearly intends to draw a connection between on the one hand Britain's national identity and imperial ambitions, symbolized here by both the lion and the flag, and on the other a by-product of coal, the fuel behind Britain's post-Industrial Revolution power.

This link between coal and both Britain's industrial power and its imperial power is not lost on the narrator of Jules Verne's *The Child of the Cavern* (1877; the original French title was *Les Indes noires* [The Black Indies]): 'We know that the English have given to their vast extent of coal-mines a very significant name. They very justly call them the "Black Indies," and these Indies have contributed perhaps even more than the Eastern Indies to swell the surprising wealth of the United Kingdom.'[40] Here the narrator clearly equates the wealth dug from Britain's coal mines with that of the vast riches plundered from the East Indies, and in so doing makes a further connection between coal and the imperial endeavour. Indeed, it could be argued that the coastal coal trade directly contributed to Britain's emergence as a major sea power and thus indirectly to the commercial and naval dominance it enjoyed by the second half of the nineteenth century. It may not be too far-fetched to speculate that just as

the Industrial Revolution could not have happened without coal, neither could the British Empire have achieved the reach it did.

Early in the twentieth century coal arrived at its zenith in Britain in terms of the number of mines producing coal, the amount of labour employed in the mines and the amount of coal produced.[41] Coal was used domestically to heat homes, and industrially to power factories and the railways in countries around the globe. Indeed, in *The Road to Wigan Pier*, Orwell insists that, 'Practically everything we do, from eating an ice to crossing the Atlantic, and from baking a loaf to writing a novel, involves the use of coal, directly or indirectly.'[42]

But in the second half of the century, the reliance on coal in many developed countries began to decline (though not in the United States, where coal use peaked in 2007[43]): people opted for cleaner forms of domestic heating, schools and factories began to replace their coal-fired boilers with oil- or gas-fired ones, and, crucially, the railways, once voracious users of coal, converted to diesel-powered locomotives. However, coal continued to fuel many industries, including the steel industry, the energy-hungry cement industry, the glass industry, the ceramic industry, the paper industry, the chemical industry and the aluminium-refining industry. Ironically, while the relative reliance on coal declined over the century, global coal use actually increased over the same period: but that consumption was largely outside the public gaze, being used

Top coal-producing countries, 2015.

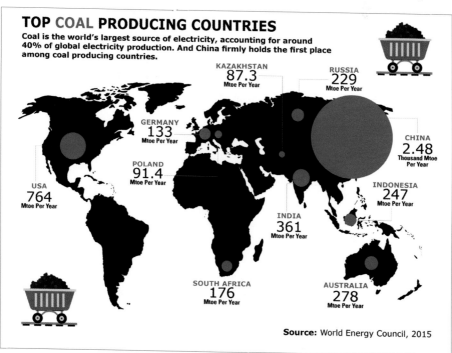

TOP COAL PRODUCING COUNTRIES

Coal is the world's largest source of electricity, accounting for around 40% of global electricity production. And China firmly holds the first place among coal producing countries.

KAZAKHSTAN
87.3
Mtoe Per Year

RUSSIA
229
Mtoe Per Year

GERMANY
133
Mtoe Per Year

CHINA
2.48
Thousand Mtoe Per Year

POLAND
91.4
Mtoe Per Year

USA
764
Mtoe Per Year

INDONESIA
247
Mtoe Per Year

INDIA
361
Mtoe Per Year

SOUTH AFRICA
176
Mtoe Per Year

AUSTRALIA
278
Mtoe Per Year

Source: World Energy Council, 2015

to produce electricity in coal-fired power plants rather than to warm domestic hearths, and, of course, the labour required to mine coal was a small fraction of what had been needed a century before.

Despite ever-growing evidence and concern about climate change, there is no sign in the twenty-first century that the world is yet ready to fundamentally turn its back on coal. Today, the most significant use of coal worldwide is in electricity generation. Electricity is fundamental to our modern way of life, providing light for homes, workplaces and streets, providing heat for both the domestic and industrial markets, and providing power for domestic appliances and industrial machinery. Currently, coal-fired power plants generate 38 per cent of global electricity.[44] More than 90 per cent of the coal mined in the United States is used in electricity generation, and although Australia exports the vast majority of the coal it mines, close to

Coal being loaded from coal mine wagons through chutes onto railway wagons at Saddle Hill, Dunedin, New Zealand, 1925.

90 per cent of its domestic coal consumption is used to generate electricity. Looked at from another angle, in round figures coal still accounts for about 30 per cent of the electricity generated in the United States, 40 per cent of the electricity generated in Germany, and a high 60 per cent of the electricity generated in Australia,[45] which has one of the highest percentages of coal-powered electricity generation in the world, but is still some way behind South Africa (77 per cent) and Poland (79 per cent).[46] And while coal-powered electricity generation in China dropped below 60 per cent of its overall energy consumption in 2018, the country's overall coal usage rose by 1 per cent.[47] Today many of us may like to think we are being environmentally conscious by choosing electricity over fossil fuels to heat our homes, but when that supposedly 'clean' power source is produced by burning

'First frost in our country, bustle in the coal industry', the Netherlands, December 1955.

brown coal, as in the Australian state of Victoria, for example, we are in fact continuing the cycle of pollution.

According to the World Coal Association, alongside electricity generation the most significant uses of coal today are in the production of steel (71 per cent of the world's steel is produced using coal), notably in India and China, who are both heavy importers of coal; in the manufacture of cement (with coal accounting for 69 per cent of the total fuel consumed in the manufacture of cement in 2017); and as a liquid fuel. Other key uses of coal continue to include the refining of aluminium, the manufacture of paper and the manufacture of chemicals and pharmaceuticals. Coal is also used in the production of activated carbon (used in water filters and kidney dialysis machines), carbon fibre (used in fishing rods and bikes), and silicone metal (used in lubricants, resins and cosmetics).[48] Despite the concentration of coal consumption in particular industries, H. Stanley Jevons's observation in *The British Coal Trade*, published in 1915, still rings true: 'There are two principal ways in which coal is used: by burning it, and by distillation', to which he added a third use: 'its application to smelting iron and other metals, where it is mixed with the molten metal in the furnace'.[49]

Today approximately two-thirds of global coal consumption is used to generate almost 40 per cent of the world's electricity, the bulk of which is expended by the wealthy countries of the world. And while the use of coal in the steel industries of China and India attracts international attention (and condemnation), the use of coal to generate the electricity that powers supposedly environmentally friendly electric trains and cars, or the electric heating that has replaced coal in homes across the Western world, tends to escape the same degree of scrutiny. Electric cars are not equal: in Australia, for example, an electric car driven in New South Wales, where more than 80 per cent of the electricity is generated by coal, is not as green as one driven in Tasmania, where about 90 per cent of electricity comes from hydro-electric generation.[50] We continue to use coal to heat homes, to power factories and to provide transport, albeit indirectly. Heat and motion – still.

Coal heap at Hanasaari power plant in Helsinki, Finland, 2010.

Interestingly, the artisanal use of coal is again proving attractive, as it did 6,000 years ago in some of the world's coal centres. In Beckley, West Virginia, home of the Beckley Exhibition Coal Mine, Appalachian coal is incorporated into hand-crafted jewellery; in Newcastle, Australia, coal gathered on the beaches is hand-faceted and made into a range of jewellery, from cufflinks to earrings, by local artist Sophia Emmett; and Silesian

Carved jet necklace, probably made in Whitby, North Yorkshire, *c.* 1875.

hard coal and silver are used to make earrings and other jewellery items in the bro.Kat studio in Katowice, Poland. It is somewhat ironic that in the twenty-first century the waste of this damaging industry is once again being used for ornament – as it was thousands of years ago, before its potential as a fuel was discovered.

3 Working the Black Seam

D. H. Lawrence begins *Sons and Lovers* (1913) with a brief history of mining in the Nottinghamshire region that provides the setting for his novel:

> 'The Bottoms' succeeded to 'Hell Row'. Hell Row was a block of thatched, bulging cottages that stood by the brook-side on Greenhill Lane. There lived the colliers who worked in the little gin-pits two fields away. The brook ran under the alder-trees, scarcely soiled by these small mines, whose coal was drawn to the surface by donkeys that plodded wearily in a circle round a gin. And all over the countryside were these same pits, some of which had been worked in the time of Charles II, the few colliers and the donkeys burrowing down like ants into the earth, making queer mounds and little black places among the corn-fields and the meadows. And the cottages of these coal-miners, in blocks and pairs here and there, together with odd farms and homes of the stockingers, straying over the parish, formed the village of Bestwood.
>
> Then, some sixty years ago, a sudden change took place. The gin-pits were elbowed aside by the large mines of the financiers. The coal and iron field of Nottinghamshire and Derbyshire was discovered. Carston, Waite and Co. appeared. Amid tremendous excitement, Lord Palmerston formally opened the company's first mine at Spinney Park, on the edge of Sherwood Forest.[1]

SURVEYING UNDER DIFFICULTIES

AT THE EDGE OF THE "GOAF"

ON THE PIT BROW—WEIGHING THE COALS

AN AWKWARD PASSAGE

The prospect described in the first paragraph, the repetition of gin-pits and donkeys slowly and monotonously circling around the gins (winches), is as bucolic as it is industrial, and exudes the sense of a scene that has remained unchanged since time immemorial. The tipping point for coal mining, signalled in the second paragraph of the quotation, came with the Industrial Revolution.

Surface mining was the earliest method of gathering coal found on or close to the surface – and the marooned colonists of Jules Verne's *The Mysterious Island* (1874) are fortunate to discover coal 'lying open to the sky',[2] which they can easily collect. Where seams were discovered on the sides of hills 'day holes' or drift mines were driven in horizontally, which provided easy ingress and egress for the colliers and the coal they dug. Shaft mining had also been practised for centuries by the time of Lawrence's description of the countryside around the village of

J. Nash,
'Work at a
Coal Mine, 1',
The Graphic,
21 September 1878.

Bestwood. If the coal lay below the surface, but not too deep, bell-pits (so-called because of their shape), were often employed. These comprised a shaft sunk to the depth of the coal seam and a winch that would be used to haul the buckets of coal to the surface. As the coal was removed the bottom of the shaft would be enlarged until in cross-section it resembled the shape of a bell. No supports were used in bell-pits, which were worked until the sides and roof of the pit became too dangerous or collapsed, and another shaft was sunk nearby.

Other early methods of mining coal underground included room-and-pillar mining and longwall mining. By far the most common of the two prior to the eighteenth century was the room-and-pillar (or pillar-and-stall) system. When the mine shaft reached the level of the coal seam a heading would be driven into the coal to open up new faces – often joining two shafts, thus improving ventilation as well as making it easier and cheaper to extract the coal. This heading would be divided into sections, each section accommodating a single hewer who would cut coal from the centre, creating a 'room' with pillars on either side to support the roof. When one side of the heading had been worked out, the hewers would start on the other side, creating a series of 'rooms' supported by pillars. An early account of room-and-pillar mining in the coalfields of northern England is found in *The Compleat Collier* (1708) by J.C.:

> After we have carried our Head-ways Drift, about eight or ten Yards from the Pit Shaft, then we consider of a Winning, how much to allow for a Winning, which is about seven Yards in these parts, or otherwise, according to the Quality, or Tenderness of it, more or less, as by Judgement is thought safest and best; out of this Winning of 7 Yards, perhaps we dare not venture to take above three Yards breadth of Coal for a Board; so that then there is but three Yards for one Man to work by himself, and therefore would be Dangerous for two Persons to Work together, least they should strike their Coal-Pics into each other, or at least hinder one another; then the Remainder of four Yards is left for a Pillar to support

the Roof and Weight of the Earth above; which makes it out so, that there is not quite half of the Coals taken out of the Ground which lies there.[3]

The size of the pillars was crucial: if the pillars were too small the roof could collapse, putting the lives of the miners at risk; if the pillars were too large, valuable coal was left behind, and the mine would be less profitable. As J.C. indicates, in order to support the roof (and also to prevent the floor pushing up) the size of the rooms and pillars were kept roughly equal, meaning that only half the available coal could be extracted.[4] The size of the pillar was determined by a range of factors, including the depth of the seam and the dip of the strata, but in general, the deeper the coalface the thicker the pillars had to be in order to support the weight above. Nonetheless, there was clearly a temptation for the mine owners to extract too much coal in order to increase their profits, leaving behind dangerously small pillars, and for the miners, who were paid piece rates, to shave extra coal from the pillars to meet quotas and/or increase their wages. One solution, as Hatcher notes, was 'the practice of systematically thinning the pillars and walls when a heading had been worked out'.[5] In the end, owners and miners alike relied on the expertise of the viewers, who were in charge of all underground activities, to judge the necessary size of pillars and keep the mine both safe and profitable.

Longwall mining, sometimes referred to as the Shropshire method, was developed in England in the late seventeenth century. It 'marked a technical advance over pillar-and-stall, and it was eventually to supersede the latter almost everywhere, because it normally resulted in far less coal being left underground'.[6] In the longwall system the miners undercut the coal across the full width of the coalface, removing coal as it fell and installing wooden props to support the roof as the coalface was pushed forward. It was, of course, best suited to locations where there was a roof of solid rock above the coal.

Regardless of whether the room-and-pillar or the longwall system of mining was employed, early mining methods were basic at best, and the extraction of coal relied on hewers who had to work in dirty, cramped conditions. The coal was hewn by hand using picks, hammers and wedges. First, lying on his side, the hewer would undercut the coal seam. He would then make a series of vertical cuts in the coalface before using his hammer and wedges to bring down the coal, working his way up from the bottom to the top of the seam. His aim was to bring the coal down in large lumps, for which he would be paid a higher rate than for small coal. As the hewers were paid piece rates their wages could vary greatly, depending on the difficulty of working particular seams or on the location they were allocated by the overman (or overlooker) who was responsible for deploying the miners. Once the hewer had brought down the coal it had to be hauled from the face to the surface. To this end the coal would first be loaded into corves (baskets made of strong hazel rods). It would then either be carried out of the pit, sometimes up a series of ladders, on the backs of young men, women or children; or the corves might be pushed or dragged on a sled by 'putters' to the pit shaft to be drawn to the surface by horses.[7] And this arduous means of carrying out the coal continued largely unchanged until the end of the seventeenth century when small wheeled barrows, carts or tubs were introduced, though they were still pushed and pulled by manual labour, first over wooden planks and then along rails. Later still, ponies were brought into the mines to pull the carts (by 1913 there were 70,000 ponies working in British coal mines; the last two were retired from a small, private drift mine in South Wales in 1999).[8] And progressively, 'black powder was introduced to blast down the coal, but undercutting, sidecutting, and drilling were still done by hand'.[9] Other gradual improvements – including James Brindley's waterwheel, installed at Wet Earth Colliery in Manchester in 1756 to pump water out of the mines, and Edward Ormerod's 'butterfly' detaching hook, a safety feature for the cages that carried miners and coal down and up the shafts – continued through the eighteenth century.

There were, however, still two major problems that needed to be addressed in the coal mines: a system to drain water from the mines; and a system to ventilate the mines, to provide fresh air, and to reduce the dangers of both firedamp (methane) and chokedamp (carbon dioxide). In drift mines where the workings rose, it was often possible to use gravity to drain the water; where the drift mines sloped down and in shaft mines water had to be lifted or pumped out. Improvements continued to be made and, as Karen Pinkus observes, 'By 1800, the British coalfields had standardized equipment and methods. The Newcomen engine pumped water out of the ground, miners used safety lamps, wrought-iron chains and sire ropes for haulage, and iron rails underground for transport.'[10] The Newcomen engine represented a major advance on earlier methods of raising water out of coal mines, replacing buckets powered by waterwheels and chain pumps (or Egyptian wheels). Writing following his visit to the north of England in 1676, Roger North notes, 'Damps or foul air kill insensibly; sinking another pit, that the air may not stagnate, is an infallible remedy.'[11] But the sinking of air shafts was by no means an adequate safeguard against damps. By 1830 'the system of coursing air through the workings by means of wooden partitions, or brattices', was well established in the northeast,[12] and thereafter various forms of power-driven ventilation were introduced – from furnace ventilation provided by a furnace at the bottom of the pit, to the much safer steam-driven devices and mechanical fans which replaced them. In *That Lass o' Lowrie's* (1877) by Frances Hodgson Burnett – better known as the author of children's books, including *Little Lord Fauntleroy* (1886) and *The Secret Garden* (1909) – 'The substitution of the mechanical fan for the old furnace at the base of the shaft, was one of the projects to which [the mining engineer] Derrick clung most tenaciously. During a two years' sojourn among the Belgian mines, he had studied the system earnestly.'[13] And, indeed, the refusal of the mine owners to act on his advice and replace their blast furnaces leads to the disaster at the centre of the novel. The other major advance for coal miners in the early nineteenth century was Sir Humphry Davy's invention of the safety lamp in 1815, which helped miners detect low oxygen levels and reduced the danger of explosions when firedamp and other flammable gases were present, as well as facilitating access to deep coal seams that might not otherwise have been accessible.

Modern methods of extracting coal have, in some respects, changed little from those used in the nineteenth century. There are still two principal methods of mining: underground mining and surface mining. The biggest change, perhaps, is in the scale of coal-mining operations, particularly where surface mining is concerned.

The main methods of surface mining today are strip mining, mountaintop removal and open-pit mining. Strip mining involves removing the soil or rock (the overburden) above the coal seam to access the coal. It is only practised in mines where the coal seam is relatively near the surface, such as the enormous Tagebau Garzweiler lignite mine – covering 11,400 hectares (114 sq. km) of Germany's North Rhine-Westphalia region – which uses the 'area-stripping' method of strip mining.[14] Conversely, 'contour mining' is used where the terrain is hilly rather than flat, and follows the contours of the hill to remove the overburden

Jack Corn, 'First Shift of Miners at the Virginia-Pocahontas Coal Company Mine #4 near Richlands, Virginia, Leaving the Elevator', April 1974.

above the coal. While both forms of strip mining have the potential to devastate the landscape, mountaintop removal, used widely across the Appalachian coalfields in recent decades, is by far the most controversial of all modern surface-mining methods. It uses explosives to remove whole mountain tops that sit above the coal seams, depositing the overburden in nearby valleys and hollows, causing major changes to the topography of the area as valleys are filled in, rivers and streams are covered, and whole ecosystems are disrupted. The controversial nature of mountaintop removal is bared in John Grisham's fast-paced thriller *Gray Mountain* (2014), a searing indictment of the big coal companies and a stark reminder that the landscape does not need to be so visibly destroyed to mine coal. Open-pit, or open-cut, mining is essentially a form of quarrying and, like strip mining, is only possible where the coal seam is close to the surface or where a number of seams can be exposed on a slope.

Underground mines continue to use either the longwall-mining method or the room-and-pillar method. Where modern mining methods do differ from earlier systems, however, is in the degree and nature of mechanization and automation employed to extract the coal. Instead of undercutting the coal seam and then ripping the coal down, as miners had done for centuries, the development of power-loading machinery allowed these

Tagebau Garzweiler surface mine, North Rhine-Westphalia, Germany.

two processes to be combined. In mines using the longwall method, a panel of coal is identified in the seam and parallel tunnels (gates) are driven along two sides of this panel, 250–350 metres apart. The tunnels are then joined by a 90-degree cut which becomes the working face. A machine known as a coal shearer scrapes the coal from the seam, and deposits the cut coal on to a loader and then on to a conveyor belt to be transported away.[15] In Britain, the biggest technical advance came with the development of the Anderton Shearer Loader, patented in 1953,

Guido Mine, Zabrze, Poland.

Longwall shearer and armoured face conveyor operating at the Twentymile underground coal mine.

which 'had a large circular drum cutter up to five feet [1.5 metres] in circumference which ran along the face, dropping the cut coal onto the integrated loader'.[16] By 1966, Anderton Shearer Loaders cut half the coal produced in British mines, and by 1977 they produced 80 per cent of the coal mined in Britain.[17] In mines where the room-and-pillar system is in use a continuous miner, a machine that extracts and loads coal at the same time, can mine as much coal in a minute as a miner in the early twentieth century could dig in a full day. While the introduction of these machines significantly increased coal production, perhaps the greatest change to coal mining in the twentieth century was that as mechanization increased, so the number of miners employed in the pits steadily decreased.

Regardless of nineteenth- and twentieth-century innovations, the conditions for miners working the black seams has remained both demanding and dangerous. Lawrence's father, Arthur (on whom Walter Morel in *Sons and Lovers* is based), born in 1846, would have worked twelve-hour shifts five days a week at the Brinsley Colliery in West Nottinghamshire from when he was first employed at the age of seven. In *D. H. Lawrence: The Early Years, 1885–1930*, John Worthen describes what awaited him when he went underground three years later:

His first job, after going underground at the age of 10, would have been opening and shutting the wind- and fire-proof doors for the passage of ponies and coal tubs. He would probably have progressed to working with the ponies; then, as a 'dayman', he would have joined a team of three or four men working under the direction of two or three 'butties' . . . The daymen – on a fixed daily wage – loaded the coal into the tubs, despatched them to the bottom of the shaft and got rid of waste material; they worked behind the 'holers', also on daily wages, who actually cut the coal by digging out the bottom of the seam to create a hole into which the overhanging coal could be broken. The butties themselves worked as supervisors and holers . . . When he was 17 or 18, Arthur would perhaps have started as a holer – the most

Miner's snap tin.

profitable job in the pit below the level of butty or manager, but also the most dangerous.[18]

Coal mining had always been a dangerous job, and not only for the holers. Miners died in roof collapses, they were killed in gas explosions, and the long hours they spent underground breathing coal dust caused serious long-term health issues. They were, in the mind of Hal Warner, the protagonist of Upton Sinclair's *King Coal* (1917), who is determined to find the truth about working conditions in the mines, 'a separate race of creatures, subterranean, gnomes' and 'stunted creatures of the dark'.[19] To meet demand for coal it was not uncommon for whole families to work down the pits, the men hewing the coal and the women and children hauling it to the surface or to the bottom of shafts from where it would be winched up. The conditions were appalling: men and women alike had to work naked or near naked because of the heat, and accidents were common. The disaster at Huskar Colliery in Silkstone, Yorkshire, which claimed the lives of 26 children in 1838, finally drove the government to address the issue of child labour in coal mines. As this disaster demonstrated all too vividly, coal mines were dangerous places, particularly for children, who were regularly injured or killed underground by explosions, by falling roofs, and also by being run over by coal carts. Many children began work underground as trappers, including Keir Hardie, a Scottish trade unionist and first leader of the British Labour Party. As William Stewart explains in his biography of the labour leader, Hardie's job as a ten-year-old working down the Newarthill pit as a trapper 'was to open and close a door which kept the air supply for the men in a given direction. It was an eerie job, all alone for ten long hours, with the underground silence only disturbed by the sighing and whistling of the air as it sought to escape through the joints of the door.'[20] There were several other jobs that children performed in the mines. Some worked as hurriers or thrusters, pulling or pushing coal tubs to and from the coalface. Hurriers would be harnessed to the coal tubs which they would then pull with help from the thrusters who would push the tubs from behind. Others assisted their fathers or older brothers by loading the coal they cut into the tubs, while some led the ponies which pulled the coal wagons. To get to their place of work underground small children often had to descend ladders and then walk large distances underground to the face. On the surface children were also employed as breakers, sorting the rock and slate from the coal with their bare hands as it moved along a belt. The fatigue caused by the combination of hard physical labour and long shifts led to high accident rates, and, of course, no compensation was paid to those who were injured.

The findings of the Children's Employment Commission (Mines) 1842 report (the first royal commission report to include illustrations) that followed the accident at Huskar Colliery shocked many people, and inspired Elizabeth Barrett Browning's poem 'The Cry of the Children' (1843), in which she encourages readers to imagine the lives of innocent children forced into manual labour: 'For, all day, we drag our burden tiring, / Through the coal-dark underground'.[21] More importantly, the report led to the Coal Mines Regulation Act of 1842. Among other things, this Act made it illegal from 1 March 1843 to employ underground any female or any boy under the age of ten. Prior to the Act, children like Lawrence's father could have worked underground as a trapper, opening and shutting ventilation doors, alone and in the dark for up to twelve hours a day, at the tender age of five or six. The Mines Act of 1842 marked the beginning of the end of child labour in Britain (other improvements included the raising of the age for boys working underground to thirteen in 1900), though there is compelling evidence that mine owners continued to employ underage children for many years, and as Peter Kirby notes in *Child Labour in Britain, 1750–1870*, 'the Mines Act tended to be applied only where it was in the interests of colliery owners'.[22]

In the United States at the start of the twentieth century, child labour laws varied from state to state, which left many children unprotected, particularly in the coal-producing regions where child labour in the mines was not considered a cause for concern. By 1908 a small number of states had introduced legislation prohibiting the employment of children in coal mines, though the minimum age varied from fourteen to fifteen to sixteen depending on the state. Kentucky specifically excluded mines from employment age limits, while Pennsylvania prohibited children under sixteen from working *inside* anthracite mines and children under fourteen from working *outside* anthracite mines (principally as breakers, sorting waste from the coal), although children as young as twelve could be employed in bituminous coal mines. It would be another thirty years before legislation was finally passed prohibiting the employment of anyone under sixteen.

Coal mining is one of the most perilous forms of child labour, yet despite international efforts to bring an end to the practice, many children as young as five still work in coal mines in countries such as India and Afghanistan. Chandrasekhar Reddy's documentary film *Fireflies in the Abyss* (2015) shines a light on the children working in Indian coal mines. As many as a fifth of all coal miners in India are believed to be children, working more than ten hours a day, descending shafts that are little more than 'rat holes', to dig coal with just a pick and head torch. Many come from Nepal or Bangladesh and are bonded to the mine owners. The conditions are hazardous and there are no safety regulations to protect them.[23] Similarly, while national laws prohibit children of any age from working in hazardous jobs such as coal mining, about a fifth of the 5,000 miners working in the hundreds of unlicensed coal mines in Afghanistan's Samangan province are underage, and working without even minimal safety standards.[24]

In Britain the Coal Mines Act of 1842 was extended in 1843 to stop all women working underground; but it did not prevent them from working on the surface. While female surface workers were found around Britain, they were particularly concentrated on the Lancashire coalfield, chiefly around Wigan, where they were known as 'pit brow lasses'. They performed a range of jobs, including screening coal at the pithead, emptying the coal tubs which came up from the face, loading the railway wagons, or tipping the loaded coal tubs into coal barges – often for less than a third of the pay of their male counterparts.[25] The women – who dressed distinctively in clogs, trousers covered by a short skirt and apron, old shirts, flannel jackets or shawls, and headscarves to keep the coal dust out of their hair – attracted considerable attention (though as John Hannavy notes, the wearing of breeches seems to have been peculiar to the women working in the Wigan pits).[26] In her first published novel, *That Lass o' Lowrie's* (1877), Frances Hodgson Burnett tells the story (much of it in dialect) of the feisty Joan Lowrie, a fictional Wigan pit brow lass. The pit brow lasses also appeared in numerous newspaper and magazine stories in the late nineteenth and early twentieth centuries, and were showcased in cabinet-card portraits and on postcards. Although there were several attempts to stop women working on the pit brows, notably in 1887 and 1911, the last pit brow lasses were not made redundant until 1972 – just as women in the United Sates were re-entering the industry, and more than a decade before women were allowed to work in coal mines in Australia.[27]

In the United States, women (including enslaved women) had worked alongside men in underground coal mines through the nineteenth century and into the twentieth before social sanctions and state laws put an end to the employment of women in coal mines. While a small number of women had worked underground in coal mines during both the First and Second

'Wigan Collieries:
Women Working
at the Coal Shoots',
cover of the *Pictorial
World*, 18 April 1874.

World Wars, legal impediments to women working as coal
miners were only removed in 1967.[28] In India, colonial reports
from the nineteenth century show that women as well as men
from the lower castes and indigenous communities worked in
Indian coal mines, and as Kuntala Lahiri-Dutt explains, they
'were part of family labour units, working as loaders who transported the coal cut by male partners from shallow open-cut
mines, or *pukuriya khads*, to the containers or tubs'.[29] International

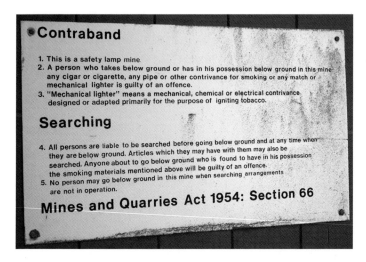

Mine regulations: signage at Big Pit National Coal Museum, Wales.

Labour Organization (ILO) conventions designed to protect women working in the coal-mining industry, adopted in 1919 and revised several times, prohibited Indian women from working in underground mines or during night shifts. It is only more recently that gender-equity legislation has been introduced in India (as elsewhere), ironically, to remove the protective conventions introduced a century earlier and allow women to work again in underground coal mines and any shifts in opencast coal mines.

The history of legislation and improved conditions in coal mines around the world has all too often been linked to pit disasters. For example, the public outcry which followed the disaster at the Hartley Pit in Northumberland – which entombed 204 men and boys on 16 January 1862, the youngest just 10, the oldest 71 – led to an Act of Parliament which spelt an end to one-shaft mines and the provision of greater support for the bereaved families of miners. The disaster is commemorated in Joseph Skipsey's poem 'The Hartley Calamity' (1862), which 'dramatizes the rescue attempt, but more importantly gives voice to the trapped coal miners'.[30]

During the course of the twentieth century, a raft of stricter safety regulations came into force in Britain, including the introduction of compulsory mines rescue stations following the 1911

Coal Mines Act, and the use of canaries in coal mines to provide an early warning of gas in the mine. The tradition of canaries in coal mines, dating from 1911, lasted until 1986, when they were replaced by electronic sensors. Similarly, a number of improvements were made to the working conditions of miners, from the increased use of underground transport to get the miners from the shaft to the coalface, to the spread of pit-head baths in the 1920s (the first was installed at Gibfield Colliery on the Lancashire coalfield in 1913).[31] Yet notwithstanding these positive changes, coal mining continued to be, and still is, a difficult and dangerous occupation around the world. Fatalities and injuries remain high, particularly in China, which now accounts for two-thirds of coal-mining deaths globally each year. A brief roll call of major twentieth-century coal-mining disasters provides a salutary lesson on the dangers of working the black seam. The worst mining disaster in Australian history occurred in 1902 at the Mount Kembla mine in New South Wales, where a gas and coal dust explosion led to the deaths of 96 miners. A royal commission determined that the explosion, triggered by a naked

Lamproom, Caphouse Colliery, National Coal Mining Museum for England.

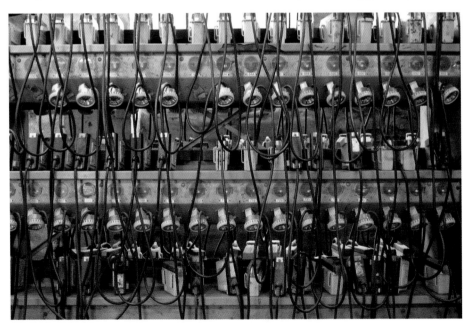

flame light, could have been avoided if safety lamps had been in use. In France a massive explosion caused by an underground fire at the Courrières coal mine in 1906 claimed the lives of 1,099 people, both underground and on the surface. The worst mining disaster in the United States occurred in 1907 when the explosions in two mines at the Monongah coal mine killed 362, many of them Italian immigrant miners. The worst mining disaster in Britain occurred at the Senghenydd Colliery in Wales in 1913, where 439 miners died following an explosion of coal dust. The following year, in 1914, an underground explosion at the Mitsubishi Hojyo coal mine in Japan led to 687 deaths. That same year the worst coal-mining disaster in Canadian history occurred at the Hillcrest Mine in Alberta, where an explosion claimed the lives of 189 miners. The explosion caused by gas and coal dust in the Benxihu Colliery in China in 1942, which claimed the lives of 1,549 miners, is the worst coal-mining disaster ever. In South Africa a huge underground collapse led to 435 deaths at the Coalbrook mine in 1960, and in the same year a methane explosion killed 684 people at the Laobaidong Colliery in China. In 1963 an explosion at the Mitsui Miike coal mine in Japan killed 458 miners and injured a further 833. In India, firedamp and coal dust explosions in two mines near Dhanbad

Pithead baths, Big Pit National Coal Museum, Wales.

killed 375 miners in 1965 and another 372 in 1975. In 1972 a series of underground explosions at the Wankie Colliery in Rhodesia (Zimbabwe) killed 426 people. And such incidents only account for a small percentage of the total mine-related deaths. The price of coal has always been high.

Some collieries were more dangerous than others. The Oaks Colliery in Hoyle Mill was considered one of the most dangerous pits in Yorkshire. Two explosions in 1846 killed a small number of miners. In 1847 a larger explosion caused by firedamp killed 73 miners. Then over two days in 1866 a series of explosions, which remains the worst mining disaster in an English coal mine, claimed the lives of 361 miners and rescuers, among them almost all the males in the village of Hoyle Mill. There were seventeen further explosions at the Oaks pit before it was closed in the 1960s.

Similarly, in Canada there were three major mining disasters at the Springhill coalfield in Nova Scotia, the first in 1891, the second in 1956 and the third just two years later in 1958. The 1891 explosion, caused by a build-up of coal dust, killed 125 miners and injured many more. The 1956 disaster was caused when derailed, loaded mine cars collided with a power cable, causing it to arc, which ignited coal dust and produced a massive explosion. A total of 39 miners, including two Drägermen (mine rescuers), died; 88 were rescued. Two years later the most severe bump, or underground earthquake, in North American history caused the collapse of sections of the No. 2 colliery, one of the deepest coal mines in the world; 74 miners died while one hundred trapped miners were brought to safety over a nine-day period.

While the vast majority of coal-mining fatalities are miners and occur underground, those who live in the shadow of the pitheads can also be at risk. The memory and the anguish of Aberfan is deeply ingrained in the minds of many Britons – and not only those in the mining communities. I remember distinctly hearing the devastating news as a nine-year-old while on a family holiday in the village of Ulpha in the Lake District's Duddon Valley. On Friday 21 October 1966 a colliery spoil tip collapsed onto the Pantglas Junior School in the South Wales mining village of Aberfan, killing 116 children and 28 adults. The danger was there for all to see, and the accident should never have happened: as Huw Morgan, the narrator of Richard Llewellyn's *How Green Was My Valley* (1939), reminisces, 'Our valley was going black, and the slag heap had grown so much it was half-way along to our house. Young I was and small I was, but young or small I knew it was wrong, and I said so to my father.'[32] His father sets aside Huw's concerns, telling him, 'That is something that will have to be done when you are grown up.'[33] The danger Huw saw in the monstrous slag heap above his village had not gone unnoticed in Aberfan, either, and numerous complaints had been directed to the National Coal Board – who, like the mine owners in Llewellyn's novel, chose to do nothing. When author Laurie Lee visited Aberfan almost a year after the tragedy he soon learned to recognize 'a village chorus rising all day from the streets and pubs, a kind of compulsive recitation of

tragedy, perpetually telling and retelling the story'.[34] Mining communities had learnt to live with death underground, but in Aberfan the roll of the dead was called in a primary school, not a mine. The Welsh writer (and son of a miner) Gwyn Thomas put this into perspective in a eulogy broadcast on the BBC Radio 4's *Today* programme:

> When Men perished in their hundreds in some eruption of blazing methane, it was possible to view it with a kind of blind ferocity, the sort of ferocity we've always used in the face of war.
>
> Men were below the earth doing a grim and unnatural job and sometimes the job would blow up in their faces and most of the doom was underground, out of sight, tucked tactfully away from public view. But Aberfan is different.[35]

Slag heaps near Garn-Yr-Erw, Wales.

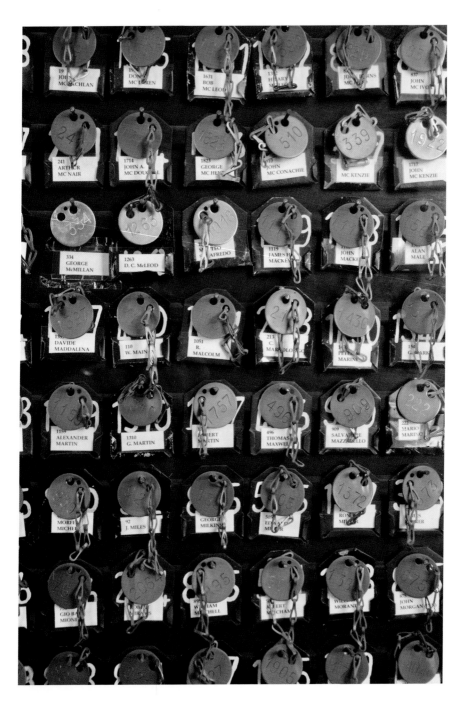

Token board,
State Coal Mine,
Wonthaggi.

Or, as Laurie Lee put it, 'Men's lives were traditionally part of the price of coal. But not those of the children, up on the surface.'[36]

Ironically, 51 giant slag heaps, the tallest rising to over 140 metres, are part of the mining landscape of the Nord-Pas de Calais Mining Basin ('a landscape shaped over three centuries of coal extraction from the 1700s to the 1900s'),[37] which was awarded world heritage status by UNESCO in 2012. It is difficult for anyone who remembers Aberfan to contemplate a slag heap, which George Orwell avers 'is at best a hideous thing, because it is so planless and functionless',[38] qualifying as a cultural achievement of 'outstanding universal value' – the precondition for world heritage listing. It is incomprehensible that any of the miners who worked the black seams around the world ever considered they might be constructing the cultural monuments of the future as they dumped the debris of their industry around the pitheads and mining villages.

Scott Morrison brandishing a lump of coal during Question Time in the House of Representatives at Parliament House in Canberra, Australia, 9 February 2017.

4 The Politics of Coal

'This is coal. Do not be afraid. Do not be scared. It will not hurt you. It is coal.'

Remarkably, these words were uttered on 9 February 2017 by Scott Morrison, then treasurer in Malcolm Turnbull's Liberal–National coalition government in Australia, as he brandished a sizeable lump of coal in the House of Representatives. His apparent intention was to taunt the Labor opposition over its desire to transition away from a reliance on coal-fired power generation. Morrison proceeded to accuse members of the opposition of 'coalophobia', which he defined as 'a malady that is affecting the jobs in the towns and industries and indeed, in this country because of the pathological, ideological opposition to coal being an important part of our sustainable and more certain energy future'.[1] Coal has long been a controversial political issue, but rarely has the matter been so hammed up.

In many countries coal mining has traditionally been associated with the fight for workers' rights. In Émile Zola's masterpiece *Germinal* (1885), the protagonist, Étienne, addressing a crowd of striking miners, urges them to continue their fight against the oppression of the mine owners: 'This time they had gone too far, the time would come when the wretched of the earth would feel they had been pushed to their limit, and they would demand justice.'[2] The key phrase, 'the wretched of the earth' (*les damnés de la terre*) – borrowed from the left-wing anthem the Internationale (1871), and later made famous by Frantz Fanon when he adopted it for the title of his 1961 book about the dehumanizing effects of colonization – screams the plight of coal miners around the world over centuries. In a similar vein, Barbara Freese argues that,

> As social outcasts who faced astonishing dangers in providing an increasingly vital commodity, [miners] also developed a fierce sense of solidarity, similar to that of soldiers in wartime. Eventually, they would come to recognize their power to band together and improve their lot in life. From this recognition would later emerge some of the strongest stirrings of the English, and American, labor movements.[3]

The politics of coal was fuelled by hardship, and played out in an ongoing, and often bitter, struggle between miners and owners. As Stephen Knight writes in his review of Greg Bogaerts's Australian mining novel *Black Diamonds and Dust* (2005), 'Mining, whether fact or fiction, had an epic simplicity. The conflict between capitalist and productive workers was never so clear: the metaphor of above and below, alive and dead, was never so literal as in the mines.'[4]

Miners' associations were in the vanguard of the trade union movement in Britain. A. J. Taylor explains that, 'Ever since the repeal of the Combination Laws in 1824 – and indeed overtly or otherwise before that time – miners had combined to assert or defend their claims and rights.'[5] By 1831, miners' unions had been established in almost every coal district in the country, but without national organization they had very little power. That came about in 1842 with the formation of the first national miners' trade union, the Miners' Association of Great Britain and Ireland. While the association only lasted seven years, Taylor suggests that it 'can justly lay claim to be the cornerstone of national miners' unionism in Great Britain'.[6] In 1889 the Miners' Federation of Great Britain (reorganized to become the National Union of Mineworkers from January 1945) was established following a national conference held in Newport, Wales, the previous year. The Miners' Federation and the union that followed it can claim to be the keystone of twentieth-century British trade unionism.

The South Wales miners' strike of 1910–11 in support of better wages and conditions – with the ensuing confrontations between striking workers and police which culminated in the Tonypandy riots in the mid-Rhondda – marks a key moment in the politics of coal. As David Smith observes, 'The very name [of Tonypandy] had come to symbolize subsequent militancy in the south Wales coalfield.'[7] Following the refusal of eighty miners to work a new seam for the wages offered, on 1 September 1910 all the miners working at the Ely Pit in Penygraig, eight hundred in total, were locked out by the owners; by early November, 12,000 miners employed by the Cambrian Combine cartel were on strike; and on the nights of 7 and 8 November striking miners rioted in the town of Tonypandy, causing widespread damage to shops and other property. But it was the decision of the then Liberal home secretary, Winston Churchill, to send troops to the Rhondda that escalated the politics of the strike and led to a degree of bitterness that still lingers in the Welsh valleys today. While the ten-month-long strike ended in defeat for the miners, it 'paved the way for an airing of the issues that resulted in the national strike of 1912 with its minimum wage provisions'.[8] Notwithstanding the Coal Mines (Minimum Wages) Act 1912, by the mid-1920s mine owners were again proposing to cut wages and extend shifts in order to maintain profits in the face of coal outputs that had been dropping steadily since 1914. Marxist historian Gwyn Williams offers a different perspective, arguing that, in South Wales at least, coal owners were still reaping the rewards of the industry, both at home and abroad:

> around the turn of the twentieth century, south Wales was one British region where growth was still breakneck and full of promise. It was Old King Coal, of course, who was the kingpin; he had conquered the British navy and his city [Cardiff] had become the greatest coal port in the world. In 1921 he had at his command 270,000 miners, with their families, one Welsh man in every four, four Welsh people in every ten.

National Union
of Mineworkers,
Edwardian
china plate.

But this was not simply a matter of coal export, huge though that was, of John Cory's bunkers straddling the world and south Wales coal keeping the greatest navy in the world afloat, staggering though these were. The capital, the technology, the enterprise, the skill and the labour of south Wales fertilized large and distant tracts of the world, from Montana and Pennsylvania to Chile, Argentina and Russia.[9]

In contrast to the well-being of Old King Coal, coal miners faced dangerous working conditions, reducing wages and increasing working hours. These would become the major grievances behind the 1926 General Strike, the only general strike in British history, called by the Trades Union Congress in support of the coal miners. Though another failure in industrial terms, the nine-day general strike did escalate the coal dispute from an industrial one to one of wider political significance. Further, it demonstrated the strength of the union movement and paved the way for the Labour Party to win the most seats in the 1929 general election, besides signalling the capacity of the coal miners to confront governments as well as colliery owners.

That power was used effectively in the 1970s under the increasingly militant leadership of the National Union of Mineworkers (NUM). In late 1971, in support of their claim for a wage rise of more than 35 per cent, miners voted for the first national strike in the coal industry since 1926. The strike, which began in January 1972, was far more successful than previous strikes, largely because of support from workers in other industries who refused to transport or unload coal, and the successful picketing of other parts of the power sector. A month after the strike began a state of emergency was declared and a three-day working week was introduced to conserve dwindling electricity supplies. By the end of February the miners were back at work, having achieved a pay rise in excess of 20 per cent, and inflicted a heavy defeat on Ted Heath's Conservative government.[10] By late 1973, however, wages for coal miners had again fallen behind those of other industries, and in February 1974 a national strike supported by every regional body of the NUM began. Heath again declared a state of emergency and implemented a three-day working week. This time, though, he also called a general election, believing public opinion was with the government and against the miners. He was wrong. The Conservatives lost their majority in the February election, and the new Labour government quickly reached an agreement with the miners that delivered them a 35 per cent pay rise followed by another 35 per cent pay rise a year later. The agreement also saw the introduction of a compensation scheme for miners who had contracted pneumoconiosis, as well as an enhanced superannuation scheme.[11]

While the strikes organized by the NUM in 1972 and 1974 were instrumental in bringing down Ted Heath's Conservative government, ten years later the miners' strike of 1984–5, the longest industrial dispute in British history, failed to do the same to the government of Margaret Thatcher. In March 1984, when the National Coal Board announced plans to close twenty pits with a loss of 20,000 jobs, miners across Britain walked out. The national strike of 1984–5 pitched miners once more into direct and deliberate conflict with the government, and there were violent clashes between picketing miners and police across the country, most infamously at the Orgreave coking plant in Yorkshire. But this time the NUM was divided, and in the absence of a national strike ballot few other trade unions supported the striking miners. Further, the government had prepared carefully for another strike – including stockpiling coal and increasing the capacity of power stations to switch from coal to oil – enabling Margaret Thatcher to engineer a crushing defeat on the NUM leadership (who may or may not have been using the strike to challenge state power) and the miners.[12] While the most obvious legacy of the strike is the destruction of the British coal industry (Britain now imports the bulk of the coal it still uses in power stations), there are other legacies that have permanently changed British society, including the use of aggressive policing, and the political defeat of the trade union movement and the working classes that prepared the way for a new era of neoliberalism.

In his poem 'Coal', Brian McCabe sets up a dialogue between the persona of the poem and the spirit of his miner father as the former goes out to the coal shed to collect coal for the evening's fire. As his father bitterly reflects on a lifetime in the mines and the solidarity of the picket lines during the 1984–5 strike, the son in turn sadly observes that all there is to show for it is 'Flattened sites, non-places, absences / surrounded by meaningless villages'.[13] The once busy pits have been replaced by derelict landscapes, and the coal communities no longer have a focus.

Dan Savage, *Strike*, 2010, screen-printed glass, Sunderland Civic Centre. Installed to mark the 25th anniversary of the end of the 1984–5 miners' strike.

Regardless of the motives of those on either side of the dispute, as McCabe's lament makes clear, there were no winners in the 1984–5 strike.

Across the Atlantic, there is a similar story. The constant reduction in wages over a number of years led directly to the coal miners' strike of 1897, organized by the United Mine Workers of America (UMWA) in the bituminous coalfields of Pennsylvania, Ohio, Indiana, Illinois and West Virginia. As J. E. George, writing in 1898, records, the wage paid to the men mining bituminous coal in the Pittsburgh district of western Pennsylvania had dropped more than 40 per cent from 1893 to 1897, and 'the condition of the miners and their families in this district was wretched':

> the average gross earnings of the miners in this district . . . were not over \$3.50 to \$4 per man per week. These conditions were made worse by the fact that the miners were compelled to deal in 'company stores,' where excessive prices were charged, and to live in company houses at excessive rents . . . What was true of Pennsylvania was true, as to wages, of Ohio, Indiana, and Illinois.[14]

The lowest wages were in West Virginia, where many of the miners were African Americans. By 1897, miners' wages were not enough to live on in any of the five states involved in the strike. The strike led to significant wage increases as well as increased membership of the UMWA. Five years later, in 1902, the UMWA called a strike in the anthracite coalfields of eastern Pennsylvania in support of the miners' claims for higher wages, shorter working days and union recognition. When the strike threatened winter fuel supplies to the major cities, President Roosevelt became involved and the federal government arbitrated a settlement that gave the miners better pay and shorter hours, while the owners got a higher price for their coal and did not have to recognize the union.

While the strikes of 1897 and 1902 (in which 'Mother' Mary Jones, the union organizer once dubbed the most dangerous woman in America, had played a key part) were largely peaceful, that changed as the twentieth century progressed.[15] Some of the most violent strikes in American coal-mining history include the West Virginia coal wars that extended from 1912 to 1921, the Colorado coalfield war of 1913–14, and the struggles in Harlan County, Kentucky, in the 1930s. The first strike in what became known as the West Virginia coal wars was the Cabin Creek and Paint Creek strike of 1912–13, where miners went on strike for better pay and conditions, and an end to the price-gouging of company stores.[16] In response the mine owners hired armed operatives of the Baldwin-Felts Detective Agency to guard their mines, drawing the battle lines that would mark almost a decade of bitter conflict. As Hoyt N. Wheeler argues, 'Firing men for union activities, beating and arresting union organizers, increasing wages to stall the union's organizational drive, and a systematic campaign of terror produced an atmosphere in which violence was inevitable.'[17] Eight years later, the 'inevitable' violence reached its zenith in the Matewan Massacre of 1920, and then the Battle of Blair Mountain in 1921, the largest labour uprising in American history, and the biggest armed insurrection since the Civil War. As Wheeler notes, what marked this conflict was not the positions of either the miners or the owners, which were in many ways predictable, but the 'out-and-out industrial warfare' that had extended over a decade and claimed numerous lives; a dispute that local union president Frank Keeney believed was an 'outright class war'.[18]

After seven months of deadly conflict that 'had already earned the Colorado coalfield war the dubious distinction of being the deadliest strike in the history of the United States',[19] on the morning of 20 April 1914 shooting broke out between a detachment of the Colorado National Guard and armed miners living in a tent colony occupied by 1,200 striking miners and their families. Accounts of who fired first are inconsistent, as are estimates of the number of deaths in what would become known as the Ludlow Massacre – venerated in folk singer and songwriter Woody Guthrie's 1946 lament of the same name. Twenty-five people died, including eleven children who burnt to death after the National Guardsmen set fire to the colony; as many as fifty

'Order Coal Now',
postcard reproduction
of First World
War poster by
Joseph Christian
Leyendecker.

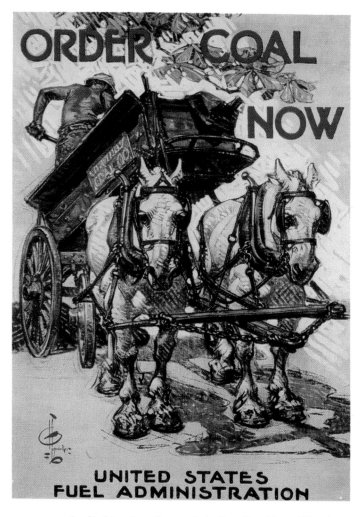

more people died in the aftermath before President Woodrow Wilson sent in federal troops to restore order.

In the 1930s, similar pitched battles were fought between miners and coal-mine operators (who often enjoyed the tacit support of local officials) in Harlan County, Kentucky. Importantly, the mine disputes in what are often tagged the 'Bloody Harlan' years focused national attention on the plight of the coal miners, while one positive outcome was the National Industrial Recovery Act of 1933, which enshrined the right of workers to

'Working Together for Victory', War Production Board poster, *c.* 1942–3.

organize unions. In her documentary *Harlan County, USA* (1976), Barbara Kopple juxtaposes the violent and bitter Brookside Mine Strike of 1973–4 with the disputes of the 1930s. As Jessica Legnini explains, 'Kopple attempts to situate the events occurring at Brookside in 1973 and 1974 as part of a historical continuum defined by antagonistic relations between mine worker and coal operator, exploitation, and poverty.'[20] It is a continuum of antagonistic relations that has driven the narrative of the politics of coal wherever the black seam has been worked. But it is a continuum that in many first-world countries is now no longer central to the politics of coal.

I REMEMBER, as a child growing up in the northern industrial town of Preston, coal-black buildings (long since scoured to reveal their true yellow sandstone and red brick facades) and smog (which in my memory descended every teatime as coal fires flared as if in unison in surrounding houses and streets). But despite the coal smoke that begrimed towns and cities across the industrialized world for centuries, it is only in recent decades that the politics of coal has turned from black to green. Whereas coal was once the cause célèbre of the left, a metonym for workers' solidarity, an instrument to hold governments to ransom, it is now more often associated with the right wing of the political spectrum; the left has turned its back on the black stuff, to embrace, instead, the environment and to advocate the urgent need to shift away from coal-generated power. This is perhaps nowhere more evident than in Australia, where pollution from Adani's proposed Carmichael Coal Mine in Queensland's Galilee Basin would lead to serious and irreversible damage to the iconic Great Barrier Reef.

Coal mining in Australia began within a decade of the arrival of the first fleet in 1788, following the discovery of black coal near the settlement that would later be named Newcastle, after the English coal port. Initially mined by convicts, coal was the first export of the fledgling colony of New South Wales. Geoscience Australia, a government organization, records that, 'Since the late 1700s about 9,100 million tonnes of black coal and about 2,300 million tonnes of brown coal have been mined and the Australian coal industry provides significant employment, capital investment and domestic and export income to the national economy.'[21] While lauding the economic benefits of coal, Geoscience Australia is, perhaps unsurprisingly, silent on the environmental impact. The economic versus environmental debates around coal in Australia have coalesced around the Carmichael Coal Mine – commonly known as the Adani Mine – since the proposed project was first announced in 2010. The Carmichael Coal Mine is a major battleground in what are shaping up to be the twenty-first century coal wars: environmentalists against the coal industry, the fight to keep coal in the ground. What happens in the Galilee Basin is likely to have consequences well beyond Australia's borders.

If it goes ahead, the proposed Carmichael Coal Mine would be the largest in Australia and one of the largest in the world. It would produce up to 60 million tonnes of mostly low-grade coal per year (2.3 billion tonnes over the sixty-year life of the mine). The coal would be transported by rail to coal ports in northern Queensland, from where it would be exported, primarily to India. In a briefing note published in 2015, the Australian Institute, a politically unaligned, independent think tank, puts the scale of the Carmichael Coal Mine into context. The mine pits would be 40 kilometres long and 10 kilometres wide, making them bigger than many capital cities; the mine's average annual emissions of carbon equivalent (CO_2-e) would be comparable to those of many countries (including Sri Lanka, Malaysia, Austria and Vietnam), and far greater than the annual emissions of many capital cities (three times those of New Delhi, six times those of Amsterdam); and the mine's output of carbon equivalent would entirely offset Australia's carbon-reduction goals. The discussion paper concludes:

> The ambition to limit global warming to no more than
> 2 degrees is inconsistent with the unfettered expansion of
> coal consumption. The vast bulk of Adani's climate impacts
> arise through the burning of its product. To maximize the
> international community's likelihood for success on reducing
> greenhouse gas emissions, Carmichael's coal product must
> remain unburnt.[22]

Nevertheless, in April 2019, Adani was granted commonwealth government approval to start building the Carmichael Coal Mine, though more approvals, at both state and federal levels, are required before coal can actually be dug out of the ground.[23]

Aerial view of the coal storage for a power plant. Gladstone, Queensland, Australia, 2009.

There are other environmental concerns, too, beyond the flouting of international attempts to limit greenhouse gas emissions. While Adani's long-term plans to build a second terminal at Abbot Point, turning it into the largest coal terminal in the world, have been temporarily scaled back, serious concerns remain for the future of the Great Barrier Reef if the expansion to the port does go ahead. Dredging will remove the sea bed and the seagrass as well as marine animals living there; the spoil will have to be dumped on land, at sea or in wetlands; increased levels of toxic coal dust will cause damage to surrounding land and marine areas, including the Great Barrier Reef; the huge increase in shipping movements will increase the risk of groundings and spills which will cause damage to wildlife.[24]

Ominously, Adani's Carmichael mine could pave the way for other large mines in the Galilee Basin. In March 2018, Waratah Coal – owned by the billionaire businessman and erstwhile politician Clive Palmer – sought approval for a huge coal mine in Central Queensland which would have the potential to produce even more coal than the Adani mine. The Alpha North Coal Mine Project would consist of series of open-cut and underground mines covering an area three times that of the Adani mine and 24 times that of Manhattan. It would have the potential to produce 80 million tonnes of coal per year, and like other Galilee Basin coal projects, would require a rail link to transport the coal to the Queensland coast. Further, the Alpha North mines could cause significant environmental damage to the Great Artesian Basin (one of the largest and deepest in the world) and to threatened species in the area, as well as having a significant impact on the Great Barrier Reef. At the time of writing, at least six other proposed mine projects for the Galilee Basin either have approval or are pending approval.[25]

In the twenty-first century the coal wars are not only about the mining of coal; they are about the use of coal. In Australia, as well as plans to open massive new mines, there are also plans to build large coal-fired power plants. Clive Palmer's Waratah Coal wants to build a 700-megawatt coal-fired power station linked to his Alpha North mine project in the Galilee Basin.[26]

Simon Denny,
Mine (a theme
park of mining
data collection, and
augmented reality),
Museum of Old and
New Art, Hobart,
June 2019–April 2020.

In March 2019 proposals for two new coal-fired power stations in New South Wales were revealed. The two 1,000-megawatt plants in the Hunter Valley would be backed by the Chinese state-owned China Energy Engineering Corporation, which has also recently built coal-fired power stations in India, Indonesia, Vietnam and Ghana. These projects, which will have a negative impact on Australia's climate-change goals, are likely to attract similar environmental opposition to that levelled at Adani's Carmichael Coal Mine.[27]

Whereas in the twentieth century the trade union movement stood in solidarity with coal miners as they battled King Coal, in the twenty-first century environmental groups are coming together to oppose Big Coal. The 'Stop Adani Alliance' is a growing network of organizations opposed to the Adani Carmichael Coal Mine, and the associated rail and port projects. Its members range from small local environmental groups to large national organizations.[28] The focus on the Adani Carmichael Coal Mine is critical to the wider concerns about the environment in general and carbon pollution in particular that are the nub of these coal wars: if the mining is stopped, the coal can't be exported or used to fuel power stations.

In a speech delivered to the anti-Adani rally in Canberra on 5 May 2019, the Tasmanian Booker Prize-winning novelist

Richard Flanagan argued that the fight against the Carmichael Coal Mine is both a fight against climate change and 'a fight for the soul of our country'. He maintains that the big coal companies are not on the side of the miners, who like their nineteenth-century forebears face pay reductions and unsafe work conditions, and argues that increased automation will effectively remove miners from the mines.[29] Big Coal in the twenty-first century, like King Coal in the nineteenth, is no friend of coal miners.

While Australia is a prime site of the coal wars, it is by no means the only site. Environmental groups across Europe are targeting coal companies over harmful emissions from coal-fired power plants. In Germany, for example, anti-coal protesters have managed to halt the expansion of the massive Hambach mine, already Europe's largest opencast mine, which would have seen the clear-felling of Hambach forest, something environmental groups see as incompatible with the country's CO_2 emissions targets. There have been similar anti-coal protests in other countries. In the Czech Republic, people rallied in 21 cities across the country in 2015 to protest against the government's plan to break its commitment to prevent any further expansion to the coal mining in North Bohemia that has destroyed nearly eighty towns since the 1960s.[30] But perhaps the most significant protests against coal are wrapped up in the huge climate-change movement initiated by Greta Thunberg and being led principally by teenage girls that has been sweeping Europe and is now expanding globally; in March 2019 protests were held in 125 countries around the world, and six months later an estimated 6 million people across 4,500 locations in 150 countries participated in the September 2019 climate strikes.[31] And the protests are having some effect as road maps are drawn up for reducing dependence on coal. Austria, Denmark, France, Finland, Italy, the Netherlands, Portugal and Britain have all signalled their intentions to phase out the use of coal completely. Germany has set a deadline of 2038 for the closure of all 84 of its coal-fired power plants.[32] While coal use may be on the wane in Western Europe, dependence remains high in Eastern Europe. Poland, which along with Germany accounts for more than half the total coal generation in the European Union, shows little inclination to reduce its dependence on coal. Indeed Poland, the Czech Republic, Bulgaria and Romania all rely on their large deposits of lignite to support the production of cheap electricity, as do some of the non-EU states in the Western Balkans. Many of these countries cannot afford to shift from coal generation to renewable-energy generation.[33] And just as many countries in the world are transitioning away from coal, others are planning to increase their use of coal. Both India and Bangladesh (which currently produces very little of its electricity from coal) have plans to massively expand their production of coal and coal-fired energy.[34] Similarly, Indonesia, already one of the largest producers and exporters of coal in the world, has significantly increased its coal production in recent years to meet increasing demand for electricity, and has no plans to reduce output in the short or medium terms. Meanwhile China continues to boost coal production to meet increasing demand for energy, despite its professed intention to reduce its dependence on fossil fuels. Between 2000 and 2009, coal production in China trebled, from 1 billion tons to 3 billion tons; at the same time imports increased as well, and in 2007 China became a net importer of coal for the first time.[35] Indeed, as Shellen Xiao Wu postulates, it is unclear whether Mao and his Cultural Revolution or coal has had the greater influence on twentieth-century Chinese history. China's coal-driven wealth and power has come at the cost of major damage to the environment and huge casualties in its mining industry.[36]

Concerns about the effects that producing and using coal have on the environment have also come to the fore in America in recent decades. Strip mines, which produced 65 per cent of the coal mined in the United States in 2017, remove the soil and rock above the coal seams; mountaintop removal, which has been increasing in West Virginia and Kentucky, changes landscapes entirely, as its name implies. And, of course, burning coal emits damaging emissions, including fly ash and bottom ash, which are by-products of using it to fuel power plants.[37] Yet, despite the enormous damage that new forms of coal extraction are causing to the environment, the administration of President Donald

Trump was committed to reviving the coal industry. However, regardless of his promise to save the country's coal industry ('Trump Digs Coal') during his election campaign, and despite his infamous withdrawal from the Paris Agreement on climate change (signed in 2016), Trump had been unable to prevent more coal-fired power plants – such as the Paradise Fossil Plant in Kentucky, run by the Tennessee Valley Authority – from closing as domestic demand for coal falters. Irrespective of Trump's promotion of what he calls 'beautiful clean coal', coal can never be clean.

It is clearly too early to declare that the war on coal is over, and that coal lost; but the era of coal, which fuelled the Industrial Revolution and has lasted for more than two hundred years, will potentially draw to a close as the world shifts inexorably towards cleaner, renewable energy. Without a shred of doubt, Donald Trump and those who share his agenda will lose this war. In 2017 Britain marked its first full day without coal power since the Industrial Revolution. In Spain, most of the country's coal mines were closed by the end of 2018. On 7 May 2019 Britain reached a new milestone, going a full week without burning coal to generate electricity, the first coal-free week in the country since the world's first coal-fired plant opened in London in 1882 (albeit made possible by nuclear power generation). Perhaps more

Coal at the side of Highway 119, near Totz, Harlan County, Kentucky.

significantly, the British government expects to have phased out coal-fired power completely by 2025.[38] The transition away from coal is both vital and inevitable, and it is progressing.

Big Pit and the surrounding landscape.

The imperative questions now are about what will happen to the vestiges of the industry: what will become of the defunct coal mines? What work will there be for the unemployed coal miners? How will the coal communities survive? In Spain, early retirement schemes for miners who are over 48, re-skilling programmes, and environmental restoration initiatives for pit communities are all part of a transition agreement that has been heralded as a model for other countries.[39] However, the afterlife for at least some of the silenced mines could lie in coal heritage and tourism.

'Loch Malcolm and Coal Town', illustration by Jules Férat for Jules Verne's
The Child of the Cavern (1877).

5 Coal Heritage and Tourism

In his 1877 novel *The Child of the Cavern*, Jules Verne anticipates
the development of coal tourism: 'Three years after the events
which have just been related, the guide-books recommended
as a "great attraction," to the numerous tourists who roam over
the county of Stirling, a visit of a few hours to the mines of
New Aberfoyle.'[1] Today, as in the time of Verne's tale, guide-
books recommend exhibition mines such as the National Mining
Museum Scotland to the many tourists who visit the country.
It is unlikely, though, that even Verne could have predicted the
growth in – and popularity of – exhibition coal mines and
museums around the globe, from Britain to Australia, and across
Europe and North America. On his comprehensive website of
caves and mines around the world, Jochen Duckeck lists well
over one hundred tourist coal mines in 27 countries across five
continents.[2] Ironically, the significant growth in coal tourist
mines over the last several decades is a direct consequence of the
decline of the coal industry worldwide: the opening of exhibition
mines, or tourist mines, has been made possible by the closure
of working mines, while coal museums now house the flotsam
and jetsam of a once vital industry.

   And whereas the beauty of the fictional New Aberfoyle
mines attracted tourists underground in Verne's novel, the
obvious attraction of coal mines today is not in spectacular
subterranean beauty (the key attraction of show caves), but in
their value as industrial and cultural heritage sites that interpret
both the technical genius of European industrialization and the
human history of the mines. Modern-day tourists observe and

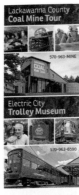

Brochures for coal tourist sites.

value the industrial scars (slag heaps, the remaining pithead infrastructure and so on) that punctuate former industrial landscapes and enter the darkness of coal mines to journey back in time, to find out how coal was extracted at different times in history, and to discover what working life was like for the miners who toiled underground. Stephen Wanhill suggests that 'what is attractive to today's tourists is not just the notion of the rural idyll but also the "rape of the fair country" that they come to see, in the form of the industrial remains that are left behind'.[3] While he is writing specifically about the Big Pit National Coal Museum in Blaenafon, South Wales, his comments could equally apply to other tourist coal-mine sites. For the descendants of coal-mining families, a visit to a tourist mine can be a secular pilgrimage motivated by a desire to learn more about their own cultural roots.

Coal tourism, a branch of the heritage tourism industry, takes a number of forms, but can usefully be divided into three main sets of attractions: underground tours, visits to colliery buildings and visits to museums. Many sites, including the Big Pit National Coal Museum, and the National Coal Mining Museum for England, located at the site of the Caphouse Colliery in Yorkshire, incorporate all three sets of attractions, as well as cafeterias and souvenir shops. Others, such as the Lancashire Mining Museum at Astley Green, offer only surface attractions in the form of colliery buildings (headgear and the engine house in the case of Astley Green). Still others, like the Kentucky Coal Museum, located in Benham, an old company mining town, are strictly what their names suggest – museums housing coalmining memorabilia. There are also replica coal mines at some sites, which in the absence of safe access to actual mines have been constructed to give visitors an impression of what it was like to work in a coal mine.

Additionally, there are industrial heritage sites, such as the River Tyne Dunston Coal Staithes, believed to be the largest wooden structure in Europe, and the recently renovated and reimagined Coal Drops Yard in London, that still have visible links with the coal industry. Other historical sites associated with the coal industry have been transformed entirely: Vienna's Kohlmarkt (Coal Market), for example, is now one of the most exclusive shopping streets in the Austrian capital.

While coal tourism is in its infancy compared to its subterranean cousin, cave tourism, some famous visitors have been guided through British coal mines over the course of the last century. George v was the first British monarch to go down a coal mine when in 1912 he descended into the Elsecar Colliery in Yorkshire, where he reportedly borrowed a miner's pick and took a turn at hewing coal. Prince Albert (later George vi), nicknamed the 'Industrial Prince', went down coal mines on several occasions, and King Amanullah of Afghanistan visited a coal mine during his trip to Britain in 1928. In 1954 Princess Margaret spent an hour underground at the Calverton Colliery in Nottinghamshire, where she emulated her grandfather in using a miner's

Miner's safety lamp carved from Welsh coal.

pick to dig out a lump of coal at the pit face.[4] In 1958 Queen
Elizabeth II descended underground when she visited the
Rothes Colliery in Scotland, while Princess Anne has been
down working mines on several occasions, including travelling
more than 3 kilometres (2 mi.) underground in 1999 to talk to
miners in Nottinghamshire's Harworth Colliery. More recently,
she opened a new underground tour at the Caphouse Colliery
in 2013.

Coal tourism, in common with other forms of industrial,
cultural or heritage tourism, is a boutique form of tourism that
offers an opportunity to preserve the industrial heritage of a
region and reinvigorate former coal-mine towns, by repurpos-
ing and managing the closed mine sites, providing work for
ex-miners (and, both directly and indirectly, others in the community), and harnessing their knowledge of a particular mine
and the coal industry more generally. It can only exist where
traces of the industry remain on the surface in the form of abandoned pithead structures and/or the underground workings can
still be accessed. Tourist coal mines are by and large former working mines that are now open to the public, usually for an entry
fee, which provide guided tours, often conducted by ex-miners
(though that is not sustainable into the future). As the guides
lead their groups through a series of shafts and galleries, past
old mine workings and abandoned machinery, they deliver an
educative commentary that may cover the geology of coal, the
history of coal mining, how coal was extracted and the working
conditions of coal miners through the ages. The tours through the
sympathetically lit passageways are often supplemented by multimedia shows, animatronics, and rides on underground trains. The
mine sites offer visitors a range of other facilities, including good
car parks, cafés, souvenir shops and even specially labelled beer.
Most tourist coal mines produce high-quality glossy brochures,
to attract tourists and in many cases to provide them with a
succinct introduction to the industrial history of the coal mine.

In the United States, the earliest 'exhibition coal mine',
opened in 1933, was not a coal mine per se, but rather the first
exhibit built at the Museum of Science and Industry in Chicago.

Coal miner statue,
Benham, Kentucky.

The very accessible tour of Old Ben 17, a mine relocated from southern Illinois, gives visitors a realistic impression of being underground in a working coal mine, and over the last 86 years it has remained popular with visitors to the museum. (Across the Atlantic, from 1952 to 1974 the very popular Mining Gallery, an 837-square-metre (9,000 sq. ft) reconstruction of a coal mine, was located in the basement of the Science Museum in London.[5]) Just five years later, in 1938, the Pocahontas Exhibition Mine in Virginia opened to visitors, making it the first exhibition coal mine in the U.S. Unusually, the tourist mine operated alongside the working parts of the mine, which continued to produce high-quality smokeless coal until 1955.

It is hard to overstate the importance of coal in the history of eastern Kentucky, and today evidence of the coal industry-cum-tourism is everywhere, from the Coal Miners Memorial Monument in Harlan to the Coal Miners Park in Benham, featuring a life-sized bronze coal miner, to the Coal House in Middlesboro, built in 1926 from 42 tons of bituminous coal. In Benham, a coal company town built by International Harvester, the Kentucky Coal Museum has a strong, albeit conservative, educational focus. Housed in the mining town's old commissary building, the museum has four floors of exhibits that portray all aspects of life in a coal camp town, as well as mining equipment and a mock mine, and a floor dedicated to country singer and coal miner's daughter Loretta Lynn. As Jessica Legnini notes, 'In recent years, as part of the attempt of Harlan County to reinvent itself as a tourist destination, the museum at Benham has taken center stage.'[6] In nearby Lynch, a town built by u.s. Steel and once one of the largest coal company towns in the country, the Portal 31 Mine, Kentucky's first exhibition coal mine, is one of the key tourist attractions in the county. In 1923 Portal 31 set a world record for the tonnage of coal brought out of a mine in a single nine-hour shift, and when it was completed in 1930 the tipple (the structure used to load coal for transport) serving Portal 31 was the biggest in the world. Today, visitors to Portal 31 travel through the drift mine on a mantrip (an underground people mover) pulled by a motor, and are greeted at several stops by very effective animatronic miners who explain the different mining methods and techniques that were used during the 46 years, from 1920 to 1966, that the mine operated. The tour concludes with a multimedia presentation that details how coal deposits were created in the Appalachian mountains, and how that coal was used to fuel the Industrial Revolution in America.

Similarly, in Pennsylvania, coal-mining heritage now has an established place in the state's tourism industry. In Lackawanna County, close to the town of Scranton, tourists can visit both the Lackawanna Coal Mine and the adjacent Anthracite Heritage Museum. The well-designed museum offers visitors a trip through

Portal 31 Exhibition Mine, Lynch, Kentucky.

Animatronic miner
and pony, Portal 31
Exhibition Mine,
Lynch, Kentucky.

the history of northeastern Pennsylvania's anthracite, or hard-coal, industry, with an informative interpretation of the exhibits and artefacts. The Lackawanna Mine Tour takes visitors through what is in many ways a carefully curated underground museum experience that is very well done, albeit a long way from the underground experience that miners faced daily. For an experience closer to that of a coal miner, however, the more rustic tour at No. 9 Mine in Lansford is as authentic as health and safety regulations will allow. Visitors are taken 490 metres (1,600 ft) into the mountain by train before continuing their tour on foot. The rough-and-ready, wonderfully eclectic and minimally curated museum above ground is packed with original equipment and holds some real treasures. The whole experience is driven by the knowledge and enthusiasm of its staff – Steve, the guide for my tour in July 2018, still works in the industry, both in a modern strip mine and underground in an independent deep mine. About 35 kilometres (22 mi.) away, the Eckley Miners' Village, administered by the Pennsylvania Historical and Museum Commission, is one of hundreds of planned company patchtowns (which provided mining families with basic needs) that were built in the anthracite region during the nineteenth century. Today comprising a number of restored miners' houses and other buildings, including two churches, the site also features a company store and breaker (where coal was cleaned and sized) which were built for the set of the film *The Molly Maguires* (1969). And, of course, coal features prominently in the tourism of other coal-mining states in the Appalachian region and beyond.

There are numerous tourist coal mines and coal heritage sites spread across Great Britain. One of the key sites is the National Coal Mining Museum for England in Yorkshire. Visitors can opt for an underground tour, which lasts for about an hour and fifteen minutes and is led by an ex-miner, who will share his experiences of working in the mine. Unusually for a tourist site, the mine adheres to strict mining safety procedures, so before entering the pit visitors are issued with hard hats and lamps, and they must surrender any battery-operated items (such as cameras, phones and watches) that could present a risk of explosion. They then descend 140 metres (460 ft) in the old cage and follow a circular route, punctuated by displays that illustrate the various ways coal was mined from the mid-nineteenth century, when whole families worked together, to the late twentieth century, when huge machines dominated the mining operations. Back on the surface there are old pit buildings to see, including the winding house, and a Clydesdale horse (that in the past would have been used for delivering coal) and several ponies (that would have worked underground) to visit in the stable yard to round off the tour. Similar safety measures are in operation at the Big Pit National Coal Museum in South Wales, which, 'As a mine . . . is subject to the full legislation of the Coal Mines Act, in which safety, as exercised by the Mines and Quarries Inspectorate, is paramount, particularly underground.'[7]

As with the National Coal Mining Museum for England, the tours are conducted by ex-miners with considerable knowledge of coal and the coal-mining industry. Above ground there are numerous historic colliery buildings to see, including the pithead baths and locker rooms, as well as exhibition spaces that recount the history of coal mining in Wales, from the mines to home life, and an excellent audio-visual presentation, *King Coal: The Mining Experience*. Close by, the Rhondda Heritage Park is located on the site of the former Lewis Merthyr Colliery, and like both the National Coal Mining Museum for England and Big Pit, it is part of the European Route of Industrial Heritage. The park offers visitors a glimpse of life in the coal communities of the Rhondda Valleys (once one of the most important coal-mining areas of the world) until the 1980s. Led by an ex-miner, the 'Black Gold Experience' includes a visit to the recently restored engine houses and the original lamp room, an underground tour of one of the colliery's mine shafts, and, with *Dram: The Cinematic Experience*, a virtual ride through the mine on the last dram (tub) of coal to the surface. Other tourist coal mines around Britain include the Apedale Coal Mine in Staffordshire, and the Hopewell Colliery in the Forest of Dean, both drift mines accessed via a

Caphouse Colliery, furnace shaft (140 m (460 ft)).

footrail or adit (a drift which slopes down to the coal seam) rather than a vertical shaft. There are also numerous tourist sites that encompass surface attractions in the form of colliery buildings and museums, including the aforementioned Lancashire Mining Museum and the Cefn Coed Colliery Museum near Neath in South Wales, which tells the story of what was once the deepest anthracite mine in the world, and one of the most dangerous in Wales in which to work.

In Europe significant efforts have been made to organize and promote coal-mine tourism, with the establishment of both pan-European and country-specific networks of coal-mine sites. The European Network of Coal Mining Museums has a representative in each of the member countries: Le Bois du Cazier in Belgium, the Centre Historique Minier in France, the Deutsches Bergbau-Museum in Germany, the Centro Italiano della Cultura del Carbone in Italy, the Muzeum Górnictwa Węglowego Zabrzu in Poland, the Museo de la Minería y de la Industria de Asturias in Spain and the National Coal Mining Museum for England in the United Kingdom. Among other scientific and

Big Pit National Coal Museum, part of the Blaenafon Industrial Landscape, a UNESCO World Heritage site.

cultural research-based objectives, the network aims to develop coal tourism and to provide visitors with an educative experience. In Belgium, the Major Mining Sites of Wallonia – comprising the Grand-Hornu, Bois-du-Luc, Bois du Cazier and Blegny-Mine – were added to the UNESCO World Heritage List in 2012. The UNESCO listing notes that the

> sites represent the best preserved places of coal mining in Belgium, from the early 19th to the second half of the 20th centuries. The Walloon Coal Basin is one of the oldest, and most emblematic of the industrial revolution, on the European continent. The four sites include numerous technical and industrial remains, relating to both the surface and the underground coal mining industry, the industrial architecture associated with the mines, worker housing, mining town urban planning and the social and human values associated with their history, in particular the memory of the Bois du Cazier disaster (1956).[8]

The Nord-Pas de Calais Mining Basin in France is also on the UNESCO World Heritage List. This landscape, which includes 87 mining villages and 51 slag heaps, as well as pits, railway stations and mining villages, which have all been turned into tourist attractions, documents all aspects of the industrial heritage of the region, from the mines themselves to the lives of the mining families who lived and worked there.[9] In Bulgaria, the Museum of Coal Mining in Pernik is situated 50 metres (165 ft) below ground in an old mining gallery, and while it is the only coal tourist site in the country, it is part of the 100 National Tourist Sites, which helps maintain its steady flow of tourists.

While there are coal tourist sites throughout Europe and Scandinavia, the most developed coal tourism industries are found in Belgium, France, Germany and Poland, where various coal mines and museums offer the tourist a full range of underground and surface attractions. At the Blegny-Mine outside Liège, for example, the tour begins with a short video about the history of coal before visitors don jackets, helmets and lamps and take the mining cage down the original shaft to the galleries located 30 and 60 metres (100 and 200 ft) underground. While all visitors must join a two-hour tour conducted in either French or Dutch, audio guides are available in English or German, making the tours accessible to a significant number of people. The underground tour is complemented by a number of surface attractions, notably the Puits-Marie (one of the oldest mining buildings in Belgium), which now houses a mining museum with exhibitions dedicated to the history of coal mining in the region, as well as the shower room, the lamp room, the electrical station and so on. There is also a restaurant that offers several Blegny Mine-branded beers.

The Centre Historique Minier (Mining History Centre) in Lewarde, which opened in 1984 on the site of the pithead of the former Delloye Colliery, is the largest mining museum in France, and a key site in the country's coal-mining tourism industry. The 8-hectare site includes numerous industrial buildings, and along with the mining museum the complex also houses the Documentary Resources Centre, which holds the archives of the Nord-Pas de Calais coalfield nationalized mining company, and the Scientific Energy Culture Centre, which explains the history of coal in the context of the wider history of energy.[10] The

Mine gallery, underground tour, Blegny-Mine, Belgium.

museum offers a number of themed exhibitions that explain the origins of coal, life in a mining village, the history of coal mining in the region, horses and mining, and the history of the Delloye Colliery. There are also guided tours of the galleries which take visitors to the coal breaker and sorting area, and explain the working conditions underground during the 270 years in which the coalfield was in operation. Audio guides to all these attractions are available in several European languages. The Puits Couriot Parc-musée de la Mine is a 3-hectare (8 ac) park surrounding the mine headframe. Focusing on the history of coal in the Saint-Étienne region, the guided tour includes a mining train and fully restored mine gallery underground, as well as a range of other heritage buildings and equipment on the surface. As Denise Cole notes, 'There has now emerged a growing appreciation, notably from the 1980s to the present, of the physical beauty of mines and mining landscapes, alongside recognition of their historical and cultural significance.'[11] Like others elsewhere, the setting of these two sites within large park-like acreages supports Cole's point that 'industrial sites have previously been victims of traditionally narrow and subjective interpretations of beauty in landscapes', and are evidence of 'the interesting juxtaposition of images that the industrial and the pastoral create'.[12]

In Germany, the Deutsches Bergbau-Museum (German Mining Museum) in Bochum is the world's largest mining museum, and one of the country's most visited museums. It currently operates four tours: 'Hard Coal', 'Mining', 'Mineral Resources' and 'Art'. The museum also offers guided tours of its reconstructed visitor's mine, 20 metres (65 ft) below the museum, and a ride to the top of the 71-metre-high (234 ft) double headframe, which previously stood above the main shaft of the Germania Mine in Dortmund-Marten, and now offers extensive views over Bochum and the Ruhr. Other key coal tourist sites in Germany include the Zollverein Coal Mine Industrial Complex in Essen, which 'consists of the complete installations of a historical coal-mining site: the pits, coking plants, railway lines, pit heaps, miner's housing and consumer and welfare facilities', which was added to the UNESCO World Heritage List in 2002.[13]

Post-industrial tourism, including coal tourism, is burgeoning in Poland, particularly in Silesia. Coal tourism in the region takes a number of forms, and demonstrates how it is possible to prevent the coal tourist experience from becoming stereotyped. The Guido Mine in Zabrze, the deepest visitor mine in Europe, offers a range of underground tours conducted in Polish (tours in English can be arranged, but they are prohibitively expensive). The mine also boasts the deepest pub in Europe, if not the world, located 320 metres (1,050 ft) underground, the final stop on each of the three tours, where Guido Mine-branded beer (as well as other refreshments and souvenirs) can be purchased.

Guido Mine, Zabrze, Poland.

At the end of their tour, visitors to the Guido Mine can enjoy a branded beer in the pub located in the pump hall, 320 m (1,050 ft) underground.

Tours at the nearby Queen Louise Adit begin on the mine railway and focus on the history of the mine and the development of mining technology over the two-hundred-year life of the mine. Also on Silesia's Industrial Monuments Route, and a short distance from the regional capital, Katowice, the beautifully preserved Nikiszowiec Settlement was built to house coal miners over a century ago. Designed by Emil and Georg Zillmann, the architecturally unique residential colony consists of four-sided three-storey red brick blocks with inner courtyards, a parish church and all the facilities required to make the community self-sufficient. Today the thriving café culture in Nikiszowiec makes it a popular destination for families as well as tourists interested in coal heritage. Nikiszowiec is also home to the Galeria Szyb Wilson (Wilson Shaft Gallery), a modern art gallery located in the old Wilson mine shaft. In Katowice itself, the Silesian Museum is built on the site of the former Katowice Coal Mine, and incorporates the mine shaft hoist tower, which has been converted into a viewing tower that provides visitors with spectacular 360-degree views over Katowice. The museum itself is spread over four underground levels, and while the museum is not a coal-mining museum, the design incorporates the tunnels, shafts and workshops of the Katowice Coal Mine.

Coal tourism takes on a different hue in Tasmania, where the Convict Coal Mines are among the numerous stops on the Australian state's convict heritage trail. The mines, located at Saltwater River on the Tasman Peninsula, near the better-known Port Arthur Historic Site, served a twofold purpose when they commenced operations in 1834: first, the mines produced coal for the Van Diemen's Land colony, limiting its dependence on coal from New South Wales; and second, they served as another means of punishing the 'worst class' of convicts. Conditions were grim in what Marcus Clarke, in his classic Australian convict novel, *For the Term of His Natural Life* (1874), called 'the dreaded Coal Mines'.[14] There were low mine headings and poor ventilation, and solitary cells deep within the underground workings to castigate those who committed further misdemeanours. Each convict had to extract 25 tons of coal in their eight-hour shift to reach the day's quota. One convict, William Thompson, transported for life for burglary in 1841, 'spent twelve months underground, harnessed with three other men to drag loaded coal carts'.[15]

For the more adventurous coal tourist, innovative new underground activities have been developed in several mines. Instead of one of the more conventional guided tours (the 'Underground Tourist Route', or the more strenuous 'Coal Mine in the Dark'), visitors to the Guido Mine can opt for the four-hour 'Miners' Shift', which offers an authentic underground work experience, followed by a shower in the pithead bath house. For younger visitors, the National Mining Museum Scotland now offers a two-hour Junior Miner Experience in which children aged eight to twelve, 'dressed as a miner and, under the watchful eye of the pit boss ... undergo training and complete a fun-filled, dirty shift at the coalface'.[16] And in what is perhaps the quirkiest coal-tourism offering, visitors to Mine 3 at Longyearbyen on the island of Spitsbergen in Norway's Svalbard archipelago have the opportunity twice a week to join a miner's workout – a training experience that utilizes old mining equipment in a ninety-minute fitness class based on the type of physical labour undertaken by miners in their daily shifts! Occasionally, working mines are open

Tea towel, State Coal Mine, Wonthaggi, Victoria, Australia.

to the interested tourist. The Coal Industry Centre in the Hunter Valley, New South Wales, for example, offers visitors 'who are seeking an enjoyable learning experience of Australia's largest export industry and domestic source of energy, a safe and fully escorted coal industry tour', as well as offering technical study tours for overseas delegations and industry partners, and educational excursions for students of all ages.[17]

Heritage tourism has initiated a new purpose for former coal mines and opened a new chapter in the story of coal. The conservation and development of decommissioned coal mines into visitor attractions has brought direct and indirect economic benefits to struggling mining communities, including employment opportunities for at least some of the former miners, who act as tour guides. Importantly, heritage tourism ensures that the cultural and industrial history of coal, and the stories of the lives of those who toiled underground, will be preserved and passed on to another generation that may know coal only as a fuel of the past. It is a coal industry for the future.

6 Coal in Literature

Coal is ubiquitous in literature. No less than fifteen of Shakespeare's plays mention coal. Coal is an important presence in the work of D. H. Lawrence, including his short story 'Odour of Chrysanthemums' (1911), in which Elizabeth Bates angrily awaits the return of her miner husband whom she thinks is out drinking, unaware that he is dead following a cave-in at the pit. Coal fuels the steam launch which Sherlock Holmes uses to chase the villains down the River Thames in Arthur Conan Doyle's *The Sign of Four* (1890), and coal is the cargo the ill-fated barque *Judea* is commissioned to carry from Newcastle to Bangkok in Joseph Conrad's autobiographical short story 'Youth: A Narrative' (1898). In Wilfred Owen's poem 'Miners', written in response to the Minnie Pit disaster, which killed 156 Staffordshire miners in 1918 (which in a letter to his mother, dated 17 January 1918, he refers to as 'the coal poem'), the narrator likens the dying miners 'writhing for air' to the deaths of soldiers in the trenches.[1] Welsh coal miners are the subject of Mervyn Peake's poem 'Rhondda Valley' (1937). In 'Reminiscences of Childhood' (1943) Dylan Thomas remembers the Wales of his childhood as 'coal-pitted', while in his famous play for voices, *Under Milk Wood* (1954), the sleeping horses in the field are 'anthracite statues'.[2] In Jennifer Maiden's poem 'Coal', the Welsh socialist politician and son of a coal miner Nye Bevan wakes up in the Lodge, in Canberra, and recognizes in then Australian prime minister Julia Gillard 'the defensive studied affability, soul / of a Welsh seaside town built on coal'.[3] Liz Berry laments the passing of the coal industry in her poem 'Black Country', while in 'Homing' she employs the vernacular

Mervyn Peake, scraperboard illustrations for his poem 'Rhonnda Valley' (1937).

of the West Midlands to tease out the connections between language and identity, class, and the pits and factories of the area: '*bibble, fittle, tay, wum*, / vowels ferrous as nails, consonants // you could lick the coal from'.[4]

The myriad references to coal in literature can be usefully partitioned into two (not necessarily exclusive) groupings: literature that *uses* coal, and literature that is *about* coal.

Literature uses coal in multiple ways: as a metaphor for poverty or suffering; to signal class; as a symbol of commerce and wealth; as a sign of pollution; as an intensifier of blackness or darkness; and, correspondingly, coal mines are used as representations of hell. For Berry, like generations of writers before her, coal is synonymous with the working classes, and commonly in literature it is employed to signal poverty and suffering. William Blake highlights the cruel fate of young boys sold into child labour in 'The Chimney Sweeper', published in *Songs of Innocence and of Experience* (1789):

> When my mother died I was very young,
> And my father sold me while yet my tongue,
> Could scarcely cry, 'weep weep weep weep'.
> So your chimneys I sweep and in soot I sleep.[5]

In Jane Austen's *Mansfield Park* (1814) the coal fire in the Price family home in Portsmouth is used to signal their lower-class status, while the socially superior Bertrams and their ilk warm themselves before wood fires. In Sean O'Casey's play *Juno and the Paycock* (1924), set amid the working-class tenements of Dublin shortly after the outbreak of the Irish Civil War in 1922, a comedic scene between Boyle and Joxer is interrupted first by the voice of a coal-block vendor chanting his wares in the street: 'Blocks ... coal-blocks! Blocks ... coal-blocks!', and then by 'the black face of the Coal Vendor' at the door.[6] And the link between coal and class is also important in Ford Madox Ford's modernist tetralogy *Parade's End* (1924–8), particularly in the later volumes, *A Man Could Stand Up* (1926) and *Last Post* (1928), which contain numerous references to coal and coal miners and to the coal disputes which followed the First World War.

In nineteenth- and early twentieth-century British literature, the polluting effects of coal are visible in both urban and rural settings. In *Hard Times* (1854), for example, Charles Dickens notes the effect of coal smoke on the red brick facades of his fictional Coketown (loosely based on the Lancashire town of Preston): 'It was a town of red brick, or of brick that would have

been red if the smoke and ashes had allowed it.'[7] The urban
landscape of A. J. Cronin's *The Stars Look Down* (1935) is equally
grim: the miners' houses in Sleescale were all 'sooty black', the
fences 'backed by heaps of slag and pit waste', and 'The far flat
background was all pit chimneys, pit heaps, pit-head gear, pit
everything.'[8] In D. H. Lawrence's *Women in Love* (1921), as the
Brangwen sisters walk away from Beldover to look at a wedding,
they pass into a rural landscape that Gudrun describes as 'like a
country in an underworld':

> The sisters were crossing a black path through a dark,
> soiled field. On the left was a large landscape, a valley
> with collieries, and opposite hills with cornfields and
> woods, all blackened with distance, as if seen through
> a veil of crepe. White and black smoke rose up in steady
> columns, magic within the dark air. Near at hand came
> the long rows of dwellings, approaching curved up the
> hill-slope, in straight lines along the brow of the hill.
> They were of darkened red brick, brittle, with dark slate
> roofs. The path on which the sisters walked was black,
> trodden-in by the feet of the recurrent colliers, and
> bounded from the field by iron fences; the stile that
> led again into the road was rubbed shiny by the
> moleskins of the passing miners.[9]

In these few sentences Lawrence repeats forms of the words
'black' or 'dark' eight times, emphasizing the way coal has polluted the once pastoral landscape, and at the same time
suggesting that the essentiality of coal and coal mining is blackness and darkness. Again, in Lawrence's *Sons and Lovers* (1913)
the green Nottinghamshire countryside is disappearing beneath
dark collieries and black slag heaps. Comparable polluted settings are found in American novels, too. In Upton Sinclair's
King Coal (1917), for example, 'There was the dry, waste grass of
the roadside, grimy with coal-dust, as was everything in the
village.'[10] In each of these novels coal is used to signify the
broader contaminated landscape.

Since the Middle Ages, literature has been thinking with coal by using it as a synonym for, and intensifier of, blackness or darkness. On a night 'dark as coal', Absalom is tricked into kissing Alison's naked arse in 'The Miller's Tale', the second of Geoffrey Chaucer's late fourteenth-century *Canterbury Tales*. The eponymous lovers of Shakespeare's narrative poem 'Venus and Adonis' (1593) are summoned to part by 'coal-black clouds that shadow heaven's light'.[11] The hero of Thomas Gray's ode 'The Descent of Odin' (1768) rides a 'coal-black steed'.[12] The ragged woman in William Wordsworth's poem 'Her Eyes are Wild' (1798) has coal-black hair, as does the Hungarian aristocrat, Countess Radsky, in Agatha Christie's *The Seven Dials Mystery* (1929), and Sir Lancelot in Alfred Lord Tennyson's 'The Lady of Shalott' (1832): 'From underneath his helmet flow'd / His coal-black curls as on he rode, / As he rode down to Camelot'.[13] R. M. Ballantyne uses coal to signal the black bodies of his native characters in *The Coral Island* (1857): the body of the cannibal chief 'was as black as coal', while a tear rolls down the 'coal-black cheek' of the Polynesian teacher.[14] Indeed, the use of coal as an intensifier of darkness is so ubiquitous that it can appear clichéd, as in Raymond Chandler's hard-boiled detective story *The Big Sleep* (1939), where the officer in the portrait which hangs in the hall of General Sherwood's palatial residence has 'hot hard coal-black eyes'.[15] However, coal still carries weight as a metaphor when used assiduously. Catherine employs the landscape of coal as a metaphor for darkness to mark the different personalities of Heathcliff and Edgar Linton in Emily Brontë's *Wuthering Heights* (1847): 'The contrast resembled what you see in exchanging a bleak, hilly, coal country for a beautiful fertile valley.'[16] In Émile Zola's *Germinal* (1885) the hewers in the Le Voreux pit work in a darkness that 'seemed to be coloured an unnatural black, with swirling waves of coal-dust'.[17] In *The Industry of Souls* (1998) Martin Booth uses coal and coal mining as a metaphor for the darkness of the soul or the dark side of human nature.

If coal is frequently a metaphor for literal or metaphorical darkness, it can also function as an enabler of warmth or comfort in literature when it is a source of fire. In Shakespeare's *Henry IV*,

Part 2 the Hostess reminds Falstaff of the promises he made sitting in her Dolphin chamber 'by a sea-coal fire', and in Byron's 'Beppo' (1818) the poet lists 'a seacoal fire, when not too dear', among the attractions of England.[18] (As well as being used to distinguish mineral coal from charcoal, the term 'sea coal' was also applied to coal washed up on the seashore or coal carried by sea, for example, from Newcastle to London.) Correspondingly, in Charlotte Brontë's *Jane Eyre* (1847) a shivering Mr Mason, newly arrived in England, 'asked for more coal to be put on the fire', and it is a handful of coal from the scuttle that fuels 'the fire's warming breath' that keeps Blair alive at the beginning of Martin Cruz Smith's *Rose* (1996).[19] In Zola's *Germinal*, even though the Grégoires' house is heated by a central boiler, 'there was a coal fire to cheer up' the dining room.[20] In George Eliot's *Silas Marner* (1861) the eponymous central character shuts the young Eppie in the coal-hole as a punishment preferable to smacking, but the three-year-old girl is unaffected by the experience, which she sees as play rather than punishment.

Coal is also used as a positive symbol of commerce – as the source of the wealth of mine owners and coal traders. George Eliot's *The Mill on the Floss* (1860) opens with an image of black ships floating on the River Floss 'laden with fresh-scented fir-planks, with rounded sacks of oil-bearing seed, or with the dark glitter of coal'.[21] Here, the adjective 'dark' exoticizes and intensifies the noun 'glitter', drawing a connection between the toil of the working-class miners, the profits of the aristocratic mine owners and trade with the rest of the world. Moreover, this scene complicates the relationship between coal and class – a complication that is also explored by other writers including Ken Follett in *Fall of Giants* (2010).

Literature further uses coal and the activities associated with coal mining as symbols of hell. In James Lee Burke's coal-mining novel *To the Bright and Shining Sun* (1971), 'the burning slag heap glowed red in the breeze'. Like the fires of perdition, 'It was a fire that never went out because its source of fuel was never stopped.'[22] In Denise Giardina's *Storming Heaven* (1987), the men who work the coke ovens built into the mountainside below the tipple at the Felco Mine 'shovelled and danced like demons at the gates of Hell', and the 'sulphurous fumes from the burning slag heap above Hunkie Holler choked the air'.[23] Equally, as he returns to the dismal coal-mining town of Wigan in *Rose*, 'The thought occurred to Blair that if Hell had a flourishing main street it would look like this.'[24] Even more vivid expressions of the coal mine as hell punctuate Zola's *Germinal*. Late in the novel, working at the coalface deep inside the mine, Catherine's vision of her fellow miners is akin to that of tormented souls in hell:

Cover design for G. A. Henty's *Facing Death: A Tale of the Coal Mines* (n.d.).

She couldn't see them clearly in the reddish glow of their lamps; they were stark naked like beasts, but so black and caked with soot and coal that she wasn't disturbed by their nudity. All you could see of their obscure labours was their spines twisting and turning like monkeys and an infernal vision of reddened limbs, toiling away amid the dull thuds and subdued groaning.[25]

The word 'infernal' directly references hell, while perdition is also evoked through the red ('reddish', 'reddened') and black ('black', 'soot', 'coal') colours which dominate this vision, and the miners themselves resemble beasts writhing and toiling in everlasting torment. Throughout the novel the colour red is used to represent the mines as hell, while coal and black signal the torment commonly associated with those residing in the netherworld.

Coal mines provide the setting for many novels, and feature prominently, for example, in a number of nineteenth-century adventure fictions, including works by Jules Verne, G. A. Henty and the American children's writer James Otis. In *The Child of the Cavern* Jules Verne revisits the territory of subterranean adventure fiction he had earlier explored in *Journey to the Centre of the Earth* (1864). The novel opens with a mysterious invitation to mining engineer James Starr to visit 'the Aberfoyle coal-mines, Dochart pit, Yarrow shaft', where 'Henry Ford, son of the old overman Simon Ford', will wait for him.[26] It turns out that Simon Ford, who has been living with his family in a cottage deep inside the mine since it was closed a decade earlier, has discovered a rich new vein of coal. Despite a cave-in, which traps them for ten days, and a series of unexplained occurrences in the mine, Starr and the Fords are not to be diverted from reopening their beloved mine, and three years later a whole town has grown up around a huge underground lake. The strange goings on at the heart of Verne's subterranean mystery, initially blamed on the supernatural, vie for space with detailed scientific and historical information about coal – including the formation of carboniferous veins, the geographical distribution of coal, its relative importance in different parts of the world, and when and where coal was first worked and used – as well as descriptions of the mining operations. As Verne explains early in the novel, 'The better to understand this narrative, it will be as well to hear a few words on the origin of coal.'[27]

The heroes of Henty's books are usually boys in their mid-teens who are plucky, confident, intensely loyal, resourceful and honourable. Jack Simpson, the hero of *Facing Death; or, The Hero of the Vaughan Pit: A Tale of the Coal Mines* (1882) has all these characteristics. Following Jack's rise from a ten-year-old gate boy in the Vaughan Pit, responsible for opening and closing his door as the coal wagons (or tubs) pass through, to 'one of the greatest authorities on mining, and the first consulting engineer, in the Black Country', the novel shows the young, working-class hero saving the mine from a strike, as well as later rescuing pitmen buried in the mine by an explosion.[28] Alongside Jack's acts of

'After the First Explosion – the Search Party', illustration by Gordon Browne for G. A. Henty's *Facing Death: A Tale of the Coal Mines* (n.d.).

derring-do, battling the unruly miners who have replaced the unruly natives of Henty's better-known imperial adventure fictions, the novel is packed with exhaustive physical descriptions of the coal mine. The detail provided following the explosion is typical:

> [Jack] knew that the danger now was not so much from the fire-damp – the explosive gas – as from the even more dreaded choke-damp, which surely follows after an explosion and the cessation of ventilation.
>
> Many more miners are killed by this choke-damp, as they hasten to the bottom of the shaft after an explosion, than by the fire itself. Choke-damp, which is carbonic acid gas, is heavier than ordinary air, and thus the lowest parts of the colliery become first filled with it, as they would with water. In all coal-mines there is a slight, sometimes a considerable, inclination, or 'dip' as it is called, of the otherwise flat bed of coal. The shaft is almost always sunk at the lower end of the area owned by the proprietors of the mine, as by this means the whole pit naturally drains to the 'sump,' or well, at the bottom of the shaft, whence it is pumped up by the engine above; the loaded wagons, too, are run down from the workings to the bottom of the shaft with comparative ease.[29]

In this and other similar passages, the wealth of accurate technical and scientific information about coal and coal mining acts as a balance to the highly implausible story Henty presents to his readers. James Otis's *Down the Slope* (1899), in which the boy-hero works as a breaker (separating impurities from the coal by hand), lacks the technical detail of Henty's novel. Nevertheless, it employs enough information about coal mines – from the noxious gases that can cause an explosion to the means of shoring up the roof of a tunnel – to bring the subterranean geography of the mine, with its numerous levels and cuttings, to life and to motor the plot to its conclusion. In each of these novels the technical descriptions of coal and coal-mining operations are intended to persuade the reader that the fictional story is underpinned by facts about coal mining.

Coal mines continue to be used as physical settings in late twentieth- and early twenty-first-century thrillers that are the heirs of nineteenth-century adventure novels. John Grisham's legal thriller *Gray Mountain* (2014) is a searing indictment of the big coal companies that are destroying the landscape of Appalachia through strip-mining, poisoning the local population with the chemicals they use in mountaintop removal, and failing to pay benefits to miners disabled with black-lung disease caused by years of breathing coal dust. Clive Cussler's sixth Isaac Bell adventure, *The Striker* (2013), is a historical thriller set in 1902 and 1912, and includes fast-paced scenes deep in the coal mines of West Virginia.

Coal is not only used as a feature of setting in literature; it is also used as a feature of character, and is commonly inscribed on the bodies of the men and women who work with coal. The titular figure of Amy Lowell's poem 'The Coal Picker' (1914) is 'Bedaubed with iridescent dirt'.[30] John Stephens, the central character of *The Earth Cries Out* (1950) by Harold C. Wells, has a black scar over one eye, which 'would always be black from the coal-dust that got in it before the gash was cleaned'.[31] In James Lee Burke's *To the Bright and Shining Sun* Big J.W.'s 'skin was grained with coal dust, rubbed so deep around the corners of his eyes that it looked like a burn', and the face of his protagonist, the sixteen-year-old third-generation miner Perry James, 'had already begun to take on the black discoloration from the coal dust'.[32] The title of Silas House's *The Coal Tattoo* (2004) refers to the 'tattoo' left by coal when it gets into cuts in the skin. In Richard Llewellyn's *How Green Was My Valley* (1939), Dai Bando's cuts are 'all dyed blue with coal dust',[33] while in Smith's *Rose*, 'Blue lines tattooed [Smallbone's] forehead. The blue was permanent, Blair knew: dust in the scars every miner collected from coal roofs.'[34] And while confronting the soldiers protecting the Le Voreux pit in Zola's *Germinal*, the striking coal miner Maheu opens his shirt to reveal 'his naked chest, with its hairy flesh tattooed with coal'.[35] The representations of coal miners that emphasize their blackened bodies mark them as other, as non-white, albeit in terms of class rather than race, and therefore as a threat to the traditional values of the establishment – in this instance the mine-owning bourgeoisie.

As the above discussion reveals, the uses of coal in literature are many and varied, and are found across the range of literary

Cover design for Harold C. Wells's *The Earth Cries Out* (1950).

Poster advertising the serialization of Émile Zola's *Germinal* in the Paris literary journal *Gil Blas* in 1884, before it appeared in book form.

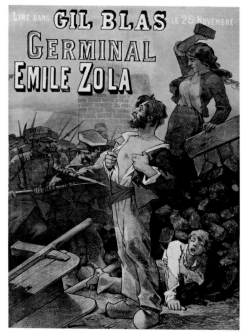

genres. Following the Industrial Revolution, however, while literature continued to use coal, coal also became the *subject* of literature. More specifically, after the Industrial Revolution, coal became a common focus for authors of the industrial novel, giving rise to a distinct body of literature – novels, short stories, poems and plays – about coal. Just as the industrial novel sought to inform readers of the plight of workers and the need for workplace reform in a broad sense, so the coal novel, which might usefully be seen as a sub-genre of the industrial novel, strives to create an awareness of the particular plight of miners and the dire working conditions they experience underground.

Coal literature has a number of patterns and tropes that define it, including detailed scientific descriptions of coal and coal-mining operations, the stygian darkness of the mines, conflict between mine owners and mine workers, a disaster in the mine, a mine rescue (not always successful) and stock settings and characters. In most coal literature, particularly the coal novel, coal is what connects the layers of the community: the bosses, the workers, the workers' families.

In Zola's *Germinal*, as in coal novels more generally, there is an interest in the hero's encounter with the subterranean environment – and coal in particular – which is invariably seen as hostile, though ultimately, as in adventure fictions set in coal mines, the true antagonists are human. In *Germinal* the miners' battles are with the mine owners who sanction the appalling working conditions they encounter underground and the dire living conditions they and their families must contend with above ground. The personification of coal and the mines as malignant or devouring beasts intensifies the threat the miners and their families face. For Brian Nelson, 'The pithead in *Germinal* is a modern figuration of the Minotaur, a monstrous beast that breathes, devours, digests and regurgitates.'[36] As the protagonist Étienne approaches the Le Voreux pit for the first time it is described as 'deep in its den, crouching like a vicious beast of prey, snorting louder and longer, as if choking on its painful digestion of human flesh'.[37] The pit is a 'nocturnal beast' with 'gaping jaws' that 'swallow[s] men down in mouthfuls of twenty or thirty at a time', 'gulping them

down effortlessly like a greedy giant'.[38] From the beginning to
the end of the novel the personification of the coal mine is consistent: it is a malevolent, man-eating beast, indifferent to the
suffering of the miners, yet subject to the control of the mine
owners.

While Zola may have been the first novelist to so dramat-
ically bring the social politics of coal to the attention of late
nineteenth-century readers, Upton Sinclair continued his on-
slaught in *King Coal* (1917), which draws heavily on details of the
Colorado coal strike of 1913–14, and the abhorrent conditions
endured by the miners and their families in the coal towns run
by America's coal barons. Sinclair's protagonist, Hal Warner, the
son of a wealthy mine owner, committed to finding out the truth
about conditions in the mining industry, adopts the alias Joe
Smith and gets a job in a mine. He soon learns that 'there is a
peculiar psychology incidental to straining your back and breaking your hands loading coal-cars in a five-foot vein; and another,
and quite different psychology, produced by living at ease off the
labours of coal-miners'.[39] Like *Germinal* and coal novels more
generally, Sinclair's novel is about men, their families and the
human price of coal. *King Coal* – like Giardina's *Storming Heaven*,
Burke's *To the Bright and Shining Sun* and numerous other novels
set in the American coalfields – also highlights another human
cost: the violence on both sides that accompanied disputes in the
American coalfields.

The lives of the mining communities are important in both
Germinal and *King Coal*, but always secondary to the task of
documenting working conditions for the miners and exposing
the cruel politics of the coal industry. The balance shifts in
Richard Llewellyn's *How Green Was My Valley*, the best-known
family saga from the coalfields. Revolving around the lives of
the Morgan family and their neighbours in a small Welsh coal-
mining village, the novel is told in flashback by son Huw as he
prepares to leave the village – the rest of his family and friends
already dead or gone. It covers a period from the late nineteenth
century to the beginning of the First World War, and is infused
with a degree of nostalgia not normally seen in coal novels.

Despite its wistful tone, the novel successfully captures the futility faced by the miners as they battle for better wages and conditions, but, like the miners in Zola's *Germinal*, are eventually forced back to work without any gain. In Llewellyn's novel, the slag heap is a 'great black bully' that whispers to itself as it advances relentlessly towards the miners' homes, while 'all the way over the mountain, slag heaps were like the backs of buried animals rising as from the Pit.'[40]

Ken Follett's huge historical novel *Fall of Giants*, the first of his Century Trilogy, also focuses on the interconnected lives of a Welsh mining community – including the mining family of Billy Williams, and the Fitzherberts, the local aristocratic family whose comforts are paid for by the miners who labour below their estates. It opens on the day of the coronation of King George v in 1911, with the minutiae of thirteen-year-old Billy Williams's first day working down the Aberowen mine:

> The pit was as dark as a night with no moon. Perhaps they did not need to see well to hew coal, Billy thought. He splashed through a puddle, and looking down saw water and mud everywhere, gleaming with the faint reflections of lamp flames. There was a strange taste in his mouth: the air was thick with coal dust. Was it possible that men breathed this all day? That must be why miners coughed and spat constantly.[41]

Like earlier coal novels, Follett's work provides his readers with a wealth of detail about coal, coal mining and the lives of the miners and their families both underground and above ground. And specifically, he uses coal to draw out issues of class in his novel. But whereas the subject of earlier writers has been exclusively coal, Follett's interest in coal lies within a broader historical context; thus the lives of his Welsh coal-mining family are intertwined with those of four other families – from England, Germany, Russia and America – in a rich historical panorama that takes the reader through the First World War and the Russian Revolution to 1924.

As in coal classics such as *Germinal* and *How Green Was My Valley*, there are also numerous underground scenes in Smith's Dickensian historical thriller *Rose*, set in the coal-mining town of Wigan, Lancashire, in 1872, where 'everything lay under a faintly glittering veil of soot.'[42] The central character, Jonathan Blair, a mining engineer turned African explorer desperate to return to 'the dark continent', reluctantly accepts a commission from Bishop Hannay, his patron and owner of the Hannay coal mines, to travel to Wigan to search for a missing curate who is the fiancé of the bishop's daughter. Alongside his tightly woven plot and adept use of Wigan dialect, Smith's descriptions of the Hannay coal mine are meticulous in their detail, including 'that curious Lancashire system of retreat, leaving a gallery of coal pillars that would slowly collapse under the weight of the earth above'.[43] Coal is everywhere in this novel. The Lancashire landscape is 'nothing but swale and hill above the long tilt of underground carboniferous forest'.[44]

Vintage postcard: 'Colliery Girl', from the collection of the National Coal Mining Museum for England.

Coal was worked into the town itself, creating coal tips that were black hills of slag. On some, coal gas escaped as little flames that darted from peak to peak like blue imps. The train slowed beside a pit as a cageload of miners reached the surface. Coated in coal dust, the men were almost invisible except for the safety lamps in their hands. The train slid past a tower topped by a headgear that, even in the subdued light, Blair saw was painted red. On the other side, figures crossed in single file across the slag, taking a shortcut home. Blair caught them in profile. They wore pants and coal dust too, but they were women.[45]

'Lancashire Pit-brow Women', *Illustrated London News*, 28 May 1887.

As in earlier coal novels, the countryside, the town, and the miners in Smith's novel all take their identity from coal. Notably, Smith's novel focuses on working women – the pit brow lasses of the Lancashire coal mines, employed to screen the coal – rather than on men and mining families. This is unusual for a novel about coal, despite the fact that, as Angela V. John notes, by the 1880s there were more than 1,300 pit brow women working in the Wigan–St Helens coalfields alone.[46] Heretofore the pit brow women had been romanticized in portraits commercially reproduced as postcards,[47] and in light-hearted poems such as 'A Pit Brow Wench for Me', published in a Wigan fortnightly journal, *The Comet*, in 1889:

> Could you but see my Nancy, among the tubs of coal,
> In tucked up skirt and breeches, she looks exceedingly
>   droll,
> Her face besmear'd with coal dust, as black as black
>   can be,
> She is a pit brow lassie but she's all the world to me.[48]

In shifting the focus of his coal novel to women and coal, Smith challenges some of the stereotypes of a male-dominated industry, and explores some of the problems the pit brow women encountered at the hands of both the male miners and the female workers who were employed in the industries, including the cotton industry, that did traditionally employ women.

Coal has been embraced on the stage as well as the page, with a number of plays focusing on coal communities having been written and performed over the years. Alan Plater's gritty documentary drama *Close the Coalhouse Door* was inspired by the stories of pitman Sid Chaplin and first performed at the Newcastle Playhouse in 1968. It uses the occasion of an old pitman's golden wedding to regale audiences with tales from the mines from the early nineteenth century to the post-nationalization present of the play, and includes several folky songs by songwriter and broadcaster Alex Glasgow, himself the son of a coal miner. Lee Hall's play *The Pitmen Painters* (2007) also has a historical perspective, looking back at coal mining in the 1930s and the story of the miners of Ashington Colliery in Northumberland who began studying art history when the local Workers' Education Association was unable to find an economics teacher.

The thirtieth anniversary of the 1984 miners' strike that transformed Britain's industrial landscape was a particularly big year for coal on the British stage. In 2014, the first of Lawrence's plays about coal miners and their families, *The Widowing of Mrs Holroyd* (1914), based on his short story 'Odour of Chrysanthemums', was revived at the Orange Tree Theatre in Richmond. The Crucible Theatre in Sheffield staged Bryony Lavery's *Queen Coal*, in which three members of a former mining family are reunited on the eve of Margaret Thatcher's funeral to celebrate. Shifting between the 2013 present and the last time the family were together almost three decades earlier, the action shows the effects of the miners' strike of 1984–5. Chris Urch's *Land of Our Fathers* (first performed in 2013), a play rich with black humour about six Welsh miners trapped underground by an electrical explosion, had a run at the Trafalgar Studios in London before embarking on a national tour. Alan Plater's *Close the Coalhouse Door* was reprised at the Oldham Coliseum. And, despite the obvious problems of re-creating a whole coal mine on stage, Beth

Steel's *Wonderland* (2014), which lays bare the grim working conditions of coal miners and confronts head-on the effects of the 1984–5 strike, was performed at the Hampstead Theatre, and had its regional premiere at the Nottingham Playhouse four years later, in 2018.

Beyond the stage, coal has also been the subject of many notable BBC radio dramas, including several set underground, where the medium's absence of a visual dimension can be used to advantage. The first play ever commissioned for radio was Richard Hughes's *A Comedy of Danger*, set in the absolute darkness of a gallery in a Welsh coal mine where three people face death after being trapped underground; originally broadcast live in 1924, it was remade and broadcast in 2013. E. Eynon Evans's *Cold Coal*, broadcast on regional radio in 1938, is set in a miner's house in the Welsh valleys and confronts the problem of massive unemployment in the coal industry. Graeme Fife's *A Misfortune at Seaham*, based on the 1880 Seaham Colliery explosion in Durham, was first broadcast in 2012. The following year Maxine Peake wrote and starred in the radio play *Queens of the Coal Age* (adapted for the stage in 2018), which dramatizes the story of

Wonderland, by Beth Steel, directed by Adam Penford, Nottingham Playhouse, 2018, production photograph by Darren Bell.

Theatre poster image for *Queens of the Coal Age* (written by Maxine Peake, directed by Bryony Shanahan, 2018).

four miners' wives who in 1993 occupy a threatened pit in an attempt to prevent further closures. As Peake's play underscores, the story of coal in literature is often inextricably linked to politics, economics and class conflict.

These central concerns are captured succinctly in Christina Rossetti's short poem for children 'A Diamond or a Coal?'. At the beginning of each stanza, an adult speaker (whom Arthur Hughes depicts as a coal monger in his illustration for the poem) poses the question 'A diamond or a coal?' to a young child.[49] In the first quatrain the child chooses a diamond, dismissing coal as 'clumsy' and worthless. But in the second quatrain, as summer gives way to winter, the child comes to appreciate the value of 'clumsy coal' over its more decorative cousin:

> A diamond or a coal?
> A diamond, if you please:
> Who cares about a clumsy coal
> Beneath the summer trees?

Arthur Hughes,
illustration for
Christina Rossetti's
'A Diamond or a
Coal?', in *Sing-song:
A Nursery Rhyme Book*
(London, 1893).

A diamond or a coal?
    A coal, sir, if you please:
One comes to care about the coal
    What time the waters freeze.[50]

Through 'the class issue implicit in the opposition between the jewel and the fuel',[51] Rossetti intimates the propensity to take coal, and by extension coal miners, for granted. It is precisely this that the corpus of coal literature seeks to expose and challenge.

7 Coal in Song and Film

As A. L. Lloyd observes in the Introduction to his compilation of ballads and songs of the coalfields, *Come All Ye Bold Miners*, 'It is doubtful whether any other industry in Britain has such a body of balladry related to the job itself, or to the life, diversions and struggles of the men engaged in that job.'[1] In other countries, too, including the United States and Canada, there are well-developed coal song canons. Universally, coal canons are dominated by protest songs that tell of the pressures facing mining communities, the solidarity among miners and their struggles against the oppression of the mine owners and managers, and, importantly, commemorate strikes and disasters.

'The Collier's Rant', a traditional Northumberland folk song which dates back to the seventeenth century, is 'the earliest known published song relating to British coal mining'.[2] 'The Coalowner and the Pitman's Wife', a strike ballad written by a Shotton Moor collier named William Hornsby at the time of the 1844 Northumberland and Durham miners' strike, is an example of the industrial folk music tradition that grew up in Britain following the Industrial Revolution, while 'Down in the Coalmine', a traditional Irish tune with surprisingly jolly words by Lancastrian songwriter J. B. Geoghegan, is a music-hall song dating from 1872 that was quickly picked up by miners in both Britain and the United States. Songs about coal continue to be numerous in the folk traditions on both sides of the Atlantic, and include several classics by Merle Travis, the son of a miner who grew up around the Kentucky coal mines. 'Sixteen Tons' (1947) is an anthem for the working miner, who risks his life in the mines

M. W. Ridley,
'Pitmen Hewing
the Coal',
The Graphic,
28 January 1871.

only to be trapped by the payment systems used in the American mines in the early twentieth century, while in the coal miner's hymn 'Dark as a Dungeon' (1947) Travis cautions young men not to seek their fortunes in the mines. Sarah Ogan Gunning's 'Come All You Coal Miners' (1937) is a call to arms, written from the perspective of a coal miner's wife, which urges miners to join the union in order to bring down the capitalist system and improve their conditions; Darrell Scott's 'You'll Never Leave Harlan Alive' (1997) tells of the exploitation of the miners by the big coal companies. Loretta Lynn's autobiographical 'Coal Miner's Daughter' (1969), one of the best-known songs in the American coal canon, describes the harsh realities of life for a miner's family in the Kentucky coalfields, while John Prine's 'Paradise' (1971) mourns the devastating impact of strip mining for coal in Muhlenberg County, Kentucky. Billy Edd Wheeler's 'Coal Tattoo' (1963) tells the story of an out-of-work miner who has nothing to show for his years in the mine except a blue tattoo on the side of his head, the result of a mining accident. Canadian singer Rita MacNeil's 'Working Man' (1982), written following a tour of the Princess Colliery, Sydney Mines, as a tribute to the miners of Nova Scotia, became an anthem for coal miners around the world.

Another coal miners' anthem, 'A Coal Miner's Life', dating from the 1890s, became a rallying song for union miners in the north of England in the 1970s and '80s. The chorus celebrates the strength of the union and the distrust towards the mine owners:

Union miners, stand together
Do not heed the owner's tale
Keep your hand upon your wages
And your eyes upon the scale

The song was famously covered by Billy Bragg (who was politicized by the miners' strike of 1984–5) on his album *The Internationale* (1990).

Like the songs of Bragg, much of the work of singer-songwriter Andy Irvine, an influential figure on the folk music scene for almost five decades now, reflects his commitment to social justice. *Precious Heroes*, his 2016 collaboration with Tasmanian Luke Plumb, 'recognise[s] some of the great, largely unknown, men and women who shaped working class and cultural roots'.[3] It includes two songs by Irvine, 'Here's a Health to Every Miner Lad' and 'Hard Times in 'Comer's Mines' – written for a concert at Dublin City University for the National Association of Mine History Organisation (NAMHO) in 2016 – which chronicle the lives of coal miners. In 'Hard Times in 'Comer's Mines', Irvine's lyrics tell the story of Nixie Boran, the communist leader of the Castlecomer miners, and detail some of the hardship the miners faced underground:

To dig out the coal you must lie on your side
For the seam is just eighteen inches high
Eight hours a day for pie in the sky
Your pay so low your expenses so high[4]

'Here's a Health to Every Miner Lad' looks at the long history of Irishmen working in coal mines, from the early days of coal mining in Pennsylvania, through to the British miners' strike of 1984–5.

The miners' strike is also the inspiration behind Sting's lament for the coal industry, 'We Work the Black Seam'. This song is unusual in the coal catalogue for the breadth with which Sting addresses his subject: it gestures to the formation of coal, to the hardship of the miners' working conditions, and to the closures of the mines in favour of nuclear power. The target of this protest song, in which Sting unambiguously positions himself alongside the miners, is unmistakably Britain's then Conservative government.

Even though many of the coal songs discussed are bitter, few can match the vitriol of the nineteenth-century folk song 'The Blackleg Miners' – another composition dating from the acrimonious 1844 miners' strike in the Northumberland and Durham coalfield – for its uncompromising attitude towards strike-breakers. The fourth verse amply demonstrates the militant attitude of the song:

Oh, don't go near the Seghill mine,
For across the mainway they hang a line,

CD cover: Andy Irvine and Luke Plumb, *Precious Heroes*, 2016.

'Helfer der Menschheit'
('Miners – Helpers of
Humanity', set of four
stamps), Deutsche
Bundespost, 1957.

To catch the throat and break the spine
Of the dirty blackleg miners.[5]

Recorded several times by Steeleye Span as 'The Blackleg Miner',
the song enjoyed a revival during the 1984–5 strike, and is used
very effectively in Beth Steel's play *Wonderland* and Richard
Lowenstein's film *Strikebound* (1984).

Just as there are numerous songs that protest the lives and
working conditions of coal miners, there are also many that
protest the deaths that are the result of those poor conditions.
Pit disasters – notably the 1958 Springhill mining disaster in
Cumberland County, Nova Scotia, which killed 74 miners, and
the 1966 Aberfan disaster in South Wales which killed 144 people,
116 of them children – have inspired many coal songs from trou-
badours on both sides of the Atlantic. Recognizing that for many
of us Aberfan is a historic milestone in our lives, the Welsh musician Martyn Joseph opens his song 'Sing to My Soul' (1998) by
asking his mother where he was on 21 October 1966; his mother
replies that he was sitting in his classroom when the black mountain of slag smothered a generation of children at a school higher

up the valley. The chorus of American songwriter Thom Parrott's 'The Aberfan Coal Tip Tragedy' (1968), written while the rescue was still under way, asks how many children died and will never grow old, and questions the human cost of coal. Fellow American singer-songwriter David Ackles was also inspired to compose a moving lament for the 116 children who died that rainy October morning in 'Aberfan' (1973). Johnny Caesar recorded a CD of fourteen coal-mining songs entitled *The Price of Coal* (n.d.), twelve of which are self-penned, including 'Senghenydd', about Britain's worst mining disaster, the 1913 pit explosion which killed 439 men and boys, and 'The Price of Coal' about the Aberfan disaster, which opens with a litany of mining accidents, and again questions the high price in human lives that we have paid for coal over the years; he also poignantly captures why this disaster, which killed a generation of children, stands apart from so many others. 'Palaces of Gold' (1968) is a cautionary conceit penned by Leon Rosselson after hearing news of the Aberfan disaster, which warns future generations that to avoid dying in the mines they should arrange to be born the sons of company directors or the daughters of judges. Other songs about the Aberfan disaster include 'A Mountain Moved' (2006) by Welsh musician John Sloman; 'Aberfan' (2003), written by American songwriter Kyle Aughe and recorded with his band Dulahan; and 'Grey October' (1968), a protest song composed by the Critics Group and Peggy Seeger, which juxtaposes what happened to the children of Aberfan with what was happening to children in the Vietnamese village of Thuy Dan during the Vietnam War (1955–75). Better known is Seeger's poignant 'The Ballad of Springhill' (1958), the most famous of all the many Springhill disaster songs, which reminds us that the price of coal can be measured in the blood and bones of miners. Or, as J. B. Geoghegan puts it, less acidly but no less effectively, in his celebration of the work of miners, 'Down in the Coalmine':

> How little do the great ones care, who sit at home secure,
> What hidden dangers colliers dare, what hardships they
>   endure;

The very fires their mansions boast to cheer themselves and
wives,
Mayhap were kindled at the cost of jovial miners lives.[6]

Coal songs are regularly showcased in films about coal.
Tracks performed by Hazel Dickens, Merle Travis, Sarah Ogan
Gunning and others provide a haunting soundtrack to Barbara
Kopple's 1976 Academy Award-winning film *Harlan County,
USA*, which documents the strike of miners at the Duke Power
Company's Brookside Mine in southeast Kentucky in 1973 and
the human toll of mining coal. Similarly, Hazel Dickens's 'Fire
in the Hole' features prominently in the soundtrack to John
Sayles's film *Matewan* (1987). In that film Sayles utilizes the tropes
of the western film genre to portray the shootout that took place
on 19 May 1920 between officials and miners of the West Virginian
community of Matewan and agents of the Baldwin-Felts agency.
Ten men died in the shootout: three locals and seven agents.
Unlike the coal novels discussed in the previous chapter, Sayles's
film does not depict the darkness of the mines, or the terrible
working conditions and ever-present danger the miners faced
underground. Instead, it stays largely above ground, focusing
on the strike and the growth of the union, and culminating in
the Matewan Massacre that, historically, can be seen as the first
shot in the biggest labour uprising in American history, which
ended in the defeat of the United Mineworkers of America in late
1921 at the Battle of Blair Mountain. (The events leading up to
and following the battle are depicted in detail in Denise Giardina's
1987 novel *Storming Heaven*.)

The Molly Maguires (1969) also highlights the violent resist-
ance to conditions in American mines in the late nineteenth
century. Set in a Pennsylvania coal town in the 1870s (and filmed
in the coal patchtown of Eckley – now administered by the
Pennsylvania Historical and Museum Commission as the Eckley
Miners' Village and Museum), Martin Ritt's film is the story of
an undercover detective hired to infiltrate and unmask the Molly
Maguires, a secret organization of Irish immigrant coal miners
fighting the exploitation and oppression meted out by the mine

The Company Store, Eckley Miners' Village, built as a movie prop for the film *The Molly Maguires* (1970).

owners. A significant amount of the film is beautifully photographed underground, detailing the full labour-intensive process of coal mining in the 1870s, from scenes of miners hewing coal with picks, to scenes showing the laden coal cars being hauled out of the mine, to a scene in a coal breaker where numerous breaker boys are at work. The low pay that went with the grim working conditions is effectively portrayed as the miners line up for their weekly pay. When the undercover detective, James McKenna, gets to the front of the line his pay is broken down as follows:

> Coal mined: 14 cars at 66c a car, total $9.24.
> Deduct: 2 kegs of powder at $2.50 a keg, $5; 2 gallons of oil at 90c a gallon, $1.80; repair 2 broken drills, 30c; pick axe, shovel, cap and lantern, $1.90
> Total deductions: $9
> Total wages for the week: 24c.

The violence of the Molly Maguires, who historically beat and murdered mine owners, supervisors and police, is balanced, though not justified, in the film by scenes like this which depict the virtual slave conditions endured by the miners in the anthracite coalfields of Pennsylvania. The poor living conditions

134

of miners and their families are also the subject of the 1933 Belgian short documentary film *Misère au Borinage*, directed by Henri Storck and Joris Ivens, which chronicles the plight of coal miners in the Borinage region of Belgium during and after a massive strike in 1932. This highly political silent film includes some excellent vintage footage of coal mining both above and below ground. It also details the cuts to miners' wages, as well as the way collieries neglected safety measures in order to reduce costs (often resulting in the deaths of miners, including two asphyxiated at the Hornu mine), and the appalling treatment of the striking miners, who are sacked and evicted from their homes.[7]

Political or militant films like *Misère au Borinage* and more recent examples, including *The Molly Maguires*, *Matewan* and the television film *Harlan County War* (2000), set in Kentucky during the 1930s, focus on the miners' struggles for better pay and conditions – and in the case of *Matewan* and *Harlan County War*, largely take place above ground. Others, including *Christmas*

Constantin Meunier, *Black Country – Borinage*, n.d., oil on canvas.

135

Miracle in Caufield, USA (1977), address the dangers miners face underground, exacerbated, in this case, by the refusal of the mine owner to close the mine, following a small explosion, in order to spray rock dust that would mitigate the risk of future potential explosions. This refusal leads to a series of explosions on Christmas Eve that traps more than sixty miners 180 metres (600 ft) below the surface. The rest of the film focuses on the race to rescue the trapped men before a bigger, more deadly explosion can occur, and the underground scenes include miners drilling and hewing coal before the explosions confine them, and the frantic work of the rescue team to reach their colleagues. The film takes on the spirit of an adventure romance, with the young, blacklisted, pro-union miner Johnny emerging as a rebel with a cause, who takes on the mantle of union organizer with a mission to make the mines safer. In a similar vein, the 1949 film *Blue Scar* (the title a nod to the coal tattoos that mark a man a miner) has a mine manager who wants his overseers to push the men to increase the mine's output at the expense of safety. The film, set in South Wales, includes all the tropes that mark the coal-mine disaster genre, including underground scenes, a roof collapse, a rescue, a death from silicosis and, like *Christmas Miracle in Caufield*, USA, the promise of improved conditions. An even earlier film, *The Stars Look Down* (1940), based on A. J. Cronin's novel set in northeast England, also focuses on the unsafe working conditions faced by coal miners, and includes underground scenes; an explosion that traps a number of miners; shots of wives, family and friends waiting anxiously on the surface; and a rescue that eventually has to be abandoned, highlighting the price of coal in miners' lives.

While many coal films open with a cage either ascending or descending a mine shaft, Richard Lowenstein's 1984 Australian film *Strikebound* opens with the flickering lamps of a group of miners emerging from the dark of an adit (a near-horizontal mine shaft). The film dramatizes the story of a coal miners' strike in the small Gippsland town of Korumburra in 1937. Based on the lives of Scottish immigrants Wattie and Agnes Doig (who appear in the opening scenes), the film traces the events that unfold at

Coal being imported to Denmark under the Marshall Plan, *c.* 1948–55.

the Sunbeam coal mine as, frustrated by poor pay and appalling conditions, the miners present the management with a list of demands. The ensuing strike/lockout, the occupation of the mine and, after six weeks, the eventual agreement to the miners' demands mark a key moment in Australia's industrial and trade union history.

The two-part British television drama *The Price of Coal* (1977), written by Barry Hines and directed by Ken Loach, is set in the fictional Milton Colliery in South Yorkshire. The first, largely comic, episode, 'Meet the People', depicts the costly cosmetic improvements being made to the mine ahead of a royal visit, at the expense of more pressing repairs. The second episode,

'Back to Reality', is much darker in tone. Infused with the dangers of coal mining, it deals with safety shortcuts that result in a fatal underground explosion and the heroism of the trapped miners and those who attempt to rescue them. The ten episodes of the American television documentary series *Coal* (2011) reveal the world of coal mining in a small West Virginia mine through the eyes of Cobalt Coal's part-owners Mike Crowder and Tom Roberts and a number of their employees. Interestingly, after the first episode aired, the Appalachia mine was cited by Mine Safety and Health Administration inspectors for activities that allegedly endangered the miners; and West Virginia's Office of Miners' Health, Safety and Training also cited Cobalt for a number of violations. The series presents what is in many ways a nostalgic portrait of the coal industry – a small company struggling to make the mine pay, employing the pillar-drilling technique to extract coal, and miners operating a continuous miner machine and working in tight spaces 180 metres underground – which is a distant cry from the reality of the massive mountaintop removal and other surface operations that produce much of America's coal today.

Whereas *Matewan* borrows the tropes of the western film genre, genre films, including thrillers and horror films, regularly employ coal mines as a setting for their narratives. At the beginning of Li Yang's Chinese psychological crime thriller *Blind Shaft* (2003), based on Liu Qingbang's novel *Sacred Woods* (*Shenmu*, 2001), a group of miners cross a desolate pithead and descend in a cage into the depths of a mine. The claustrophobic scenes in the mine are the prelude to a collapse, but not the usual tragedy caused by safety shortcuts. In this film the cave-in is deliberately triggered by a miner after he has murdered his younger colleague with a pickaxe. His partner then lodges a claim for compensation, claiming that the miner who died in the 'accident' is his brother. This, it transpires, is the modus operandi for the scam the two con artists have been practising around the country: moving from one mine to another, they pick up a naive job-seeker whom they pass off as a relative, murder him in the mine, then collect a payout from the mine owners who are desperate to avoid

LMS BRITISH INDUSTRIES
COAL
BY G. CLAUSEN, R.A.

LMS poster for British Industries, by George Clausen, RA.

an official inquiry. Shot largely underground, the film is about darkness and the loss of humanity, and beyond its success as a noir thriller it offers a devastating exposé of the conditions faced by miners in China's deadly, and often illegal and unregulated, coal industry. In coal films, miners are usually depicted as a community of men toiling together in the face of brutal conditions and heartless pit owners; consequently, the discovery that the two miners are unscrupulous con artists and brutal murderers is particularly shocking and unexpected.

Coal mines have also provided the setting for at least one horror film. Ben Ketai sets his low-budget film *Beneath* (2013) in a coal mine where a group of miners become trapped far underground following a collapse. While the relationship between the retiring miner, George, and his daughter Sam, an environmental lawyer, offers a missed opportunity to develop an interesting storyline based on the politics of coal, the film does contain some noteworthy early scenes showing underground operations in a coal mine, including a continuous miner – a machine equipped with tungsten carbide teeth that rips coal from the seam – at work. Somewhat predictably, the continuous miner later starts without explanation to chew up one of the trapped miners as the crew quickly and unconvincingly descend into madness, and the supernatural horror elements of the film push aside the more realistic elements of a mining-disaster narrative.

Perhaps more surprisingly, coal and the mining industry have provided the backdrop to a number of successful British kitchen sink dramas including *Brassed Off* (1996), *Billy Elliot* (2000) and *Pride* (2014). Based loosely on the story of the Grimethorpe Colliery Band, Mark Herman's *Brassed Off* is set during the wave of pit closures that preceded the privatization of British Coal in the mid-1990s. Apart from following the trials and tribulations of the Yorkshire colliery band and its various members, the film also references the high number of suicides that followed the closures of the pits, and the debts that some miners still faced from the long 1984–5 strike which may have prompted them to accept the relatively generous redundancies on offer a decade later. Indeed, while the emotional story of the band's success in the National Brass Band Championships and the rekindled romance between local miner Andy Barrow and Gloria Mullins, who has returned to the fictional Grimley to report on the profitability of the pit, are key to the success of the film as a romantic comedy-drama, the politics of the coal industry are never far from the surface.

Stephen Daldry's *Billy Elliot*, set in County Durham against the backdrop of the 1984–5 coal miners' strike, foregrounds the story of a miner's son who wants to be a ballet dancer. Both the pithead scenes during the strike, and the scene towards the end of the film where a group of miners descend into the pit following the collapse of the strike (filmed at Ellington Colliery, the last operational deep coal mine in northeast England; it closed in 2005), as well as the frequent scenes where striking miners clash with the omnipresent police, provide a degree of gritty realism about life in a tight-knit coal community during a period of massive industrial upheaval. In many ways, the film is as much about the loss of community as it is about a boy who wants to dance, or the miners' strike. Yet it is coal that holds the community together, and it is ever-present in the film – whether it is the dance teacher Mrs Wilkinson's husband pontificating on the economics of coal, or a school lesson where the teacher is explaining the formation of coal, or the looming background presence of the pithead when Billy and his father visit his mother's grave. Like *Billy Elliot*, the historically based comedy-drama *Pride*, directed by Matthew Warchus, is also set against the British miners' strike of 1984–5. While the film is concerned with a range of social issues – including the impact of HIV, the age of consent for gay men, and hostility towards gay men and lesbians – the group that calls itself Lesbians and Gays Support the Miners (LGSM) balances the film's focus evenly between gay issues and the miners' strike by concentrating on the unlikely alliance that was forged between the small group of activists that made up LGSM and mining families in the small South Wales village of Onllwyn. Like *Billy Elliot* and *Brassed Off*, *Pride* focuses on community; coal is the 'dark artery' that brings together the disparate gay and mining communities in the face of Thatcherism, just as it connects the mines of South Wales and Pennsylvania, as the older miner Cliff describes with great pride in the film:

> It's called the Great Atlantic Fault. (*Traces along the bar*)
> And it starts here. In Spain. And it goes under the Bay
> of Biscay. Then it comes up in South Wales. Then it goes
> under the Atlantic for miles and miles. Then it comes up
> again in Pennsylvania . . . You could take a miner from
> Wales or Spain or America and show them that seam
> and they would recognise it. There's no other coal like it.
> It's perfect. Pure.[8]

The nostalgia that marks Cliff's reflection runs through the canons of both coal songs and coal films. Songs and films alike are

about humble heroes who risked their lives crawling underground to dig the coal that fuelled the Industrial Revolution and provided heating and lighting for homes around the world; they recount the dangers the miners faced every time they descended into a coal mine; and they seethe with anger at the exploitation the miners and their families suffered at the hands of unscrupulous mine owners.

Coal seam, Lackawanna Coal Mine tour, Scranton, Pennsylvania.

8 Coal in Art

Coal, the coal industry and coal communities have attracted the interest of artists over several centuries – from landscapes featuring coal mines to depictions of miners working at the coalface, and from the distribution and myriad uses of coal to the everyday lives of coal communities.

In 1982–3 the Arts Council of Great Britain and the National Coal Board organized a touring exhibition that visited Stoke-on-Trent, Swansea, London, Durham and Nottingham. *Coal: British Mining in Art, 1680–1980* brought together three centuries of art that shared the common subject of coal, and showed how coal art has shifted in terms of its subject matter over the years. The earliest work included in the exhibition was Peter Hartover's *Coal Staithes on the River Wear and Lumley Castle in the Distance* (1680). On first viewing, the coal staithes and colliers appear less the subject of the painting than a part of the larger narrative scene. Closer scrutiny, however, reveals that coal is both at the centre of the canvas and its central subject: as Douglas Gray explains, 'The whole scene and activity therein depends on coal, its mining, its export, and its sale. Not one of the human pursuits portrayed is possible without it.'[1] Representations of actual coal mines came later, towards the end of the eighteenth century, with works such as George Robertson's *A View of the Mouth of a Coal Pit near Broseley, in Shropshire* (1788), Julius Caesar Ibbetson's *Coal Mine* (c. 1790) and John Hassell's *A View near Neath in Glamorganshire, South Wales* (1798). And from the early nineteenth century onwards, artworks unsentimentally representing coal mining and coal

George Robertson,
*A View of the Mouth of
a Coal Pit near Brosely,
in Shropshire*, 1788,
etching and engraving.

miners – including George Walker's *Yorkshire Miner* (1814) – became more widespread.

Coal and coal mines became more discernibly the subjects of paintings in the early nineteenth century as many leading artists turned their attention to coal pictures. Works like T. M. Richardson's *North Eastern Colliery, Murton* (1841) and John Wilson Carmichael's *A View of Murton Colliery near Seaham, County Durham* (1843) show the impact of the discovery of coal and the subsequent sinking of a mine on what had been a tiny hamlet. Henry Perlee Parker created a series of paintings of miners between 1836 and 1853, including *Pitmen at Play*, exhibited at the Royal Academy in 1836. Constantin Meunier produced numerous coal works in various mediums following his visit to the Belgian coalfields in 1881. Many of these works focus on the miners themselves, including the paintings *The Return of the Coal Miners* (*c.* 1881–9) and *Return from the Mine* (*c.* 1881–9), as well as a large number of works in bronze, including a life-size statue, *The Crouching Miner* (1903); the relief *The Mine (Monument to Labour)* (1901), which shows miners working underground; and one of his most celebrated works, *Firedamp* (1889), which captures the moment a mother recognizes her son among the dead in the aftermath of a mining accident. Other paintings by Meunier such as *In the Black Country* (1893) and *Black Country*

– *Borinage* (*c.* 1893) show the impact of industrial coal mining on the landscape. *In the Black Country* is dominated by slag heaps and mine buildings, with dark chimneys belching out thick black smoke which all but obscures what little of the countryside can be seen in the distance. Similarly, L. S. Lowry, best known for his depictions of industrial landscapes in his native Lancashire, captures the dour impact of a mine on the countryside in his drawing *Wet Earth Colliery, Dixon Fold* (1924), one of several collieries that were close to his home in Pendlebury.

The transportation of coal, rather than the mines themselves, attracted the attention of some of the best-known artists of the nineteenth century, including J.M.W. Turner, Claude Monet and Vincent van Gogh. Turner exhibited *Keelmen Heaving in Coals by Moonlight*, which illustrates the importance of coal to the Industrial Revolution, at the Royal Academy in 1835. The keelmen, silhouetted against the flames that light their toil, are shown transferring coal from the flat-bottomed keels that have brought it down the River Tyne from the coalfields of North-umberland and Durham onto ocean-going sailing ships that will transport it to London, where it will feed the voracious appetites of factories like those belching smoke in the shadowy distance of the painting. *The Coalmen* (1875) is unusual among Monet's works in its focus on labourers. The lower left of this dark painting is

George Walker,
Yorkshire Miner,
1814, lithograph.

John Wilson
Carmichael, *A View
of Murton Colliery
near Seaham, County
Durham*, 1843,
oil on canvas.

J.M.W. Turner,
*Keelmen Heaving in
Coals by Moonlight*,
1835, oil on canvas.

dominated by a line of barges laden with coal from the mines of
northern France or Belgium which both draws the viewer's attention and, cutting diagonally across the canvas, guides his or her
gaze under the Pont d'Asnières to the factories whose chimneys
spew smoke in the hazy background. On the barges, men work
in the holds filling baskets with coal, while columns of shadowy,
ant-like workers monotonously haul the heavy baskets up the
precarious wooden gangplanks and return down the narrow
walkways with empty containers on their shoulders. The work is
dull, dirty and back-breaking, and as Mike McKiernan notes, in
nineteenth-century Paris, 'a typical 300 tonne barge might take
a week to unload'.[2] And though he may not have intended his
painting to be an exercise in social criticism, Monet does nevertheless emphasize 'the bleakness of the workers' plight' as they
go about the business of unloading the coal barges.[3] Coal barges
are also the subject of two paintings by Van Gogh: *Coal Barges*
(1888) and *The Stevedores in Arles* (1888). In both paintings the coal
barges and the coal stevedores unloading them are silhouetted
against the orange/yellow sunsets, emphasizing the blackness of
the coal and the dirtiness of the work.

Claude Monet,
The Coalmen, 1875,
oil on canvas.

Constantin Meunier, *Triptych of the Mine (Descent, Calvary, Return)*, 1902, oil on canvas.

Van Gogh's sympathy for the working classes is also evident in another, quite different, coal painting, *Miners' Wives Carrying Sacks of Coal* (1882), where the women are scarcely distinguishable from the heavy coal sacks they carry on their shoulders. Indeed, coal art increasingly focused on miners, their families and mining communities. Like Van Gogh's coal paintings, Constantin Meunier's *Triptych of the Mine* (1902) depicts miners as heroic peasants worn down by their labours. *Calvary*, the central panel of Meunier's triptych, shows miners trudging to work under a smoke-filled sky; the left panel, *Descent*, shows miners entering the cage that will take them down into the darkness of the mine at the beginning of their shift; while the right panel, *Return*, shows the miners emerging from the lift shaft, worn out at the end of their long shift. As Linda Nochlin explains, 'Constantin Meunier's *Triptych of the Mine* transformed the traditional motif of the Road to Calvary into a poignant visual chronicle of the mine worker's brutalizing occupation.'[4]

Artists also gradually turned from the landscape and depictions of miners above ground to the underground landscape and the daily toil of the miners at the coalface. An early underground scene in Thomas H. Hair's watercolour *Bottom of the Shaft, Walbottle Colliery* (1842) shows baskets of coal being loaded onto trams, ready to be pulled by ponies to the bottom of the pit shaft from where they would be hauled to the surface, while in *The Miners* (1917) the Anglo-Welsh artist Sir Frank Brangwyn

Vincent van Gogh, *Coal Barges*, 1888, oil on canvas.

Vincent van Gogh, *Miners' Wives Carrying Sacks of Coal*, 1882, oil on canvas.

149

– who painted a *Self-portrait with Miners* (1907) – depicts an underground scene with miners pushing trucks of coal. The work of German artist Hermann Kätelhön includes an important portfolio of etchings from the 1930s that records the daily lives of miners and their work underground. In the following decade, as a war artist during the Second World War, Henry Moore produced several sketches of miners working at the coalface, including *Men Clearing Coal Face and Climbing over Belt* (1942), after spending time underground in the collieries of his native Yorkshire. The coal art of erstwhile Derbyshire miner George Bissell (who was also commissioned to produce images of coal miners at work during the Second World War) evocatively captures the visceral and claustrophobic nature of work underground where stylized and physically large men crouch awkwardly in the cramped galleries, their muscles straining with the effort of their labour – cutting, wringing or heaving coal, timbering or removing props. The aftermath of an underground explosion is captured in *After the Blast* (c. 1935) by Vincent Evans, another ex-coal miner and artist known for his mining subjects. The work of Australian artist and social campaigner Noel Counihan, who spent time in the Wonthaggi coalfields in 1944 sketching miners at work, includes *In the 18 Inch Seam, State Coal Mine, Wonthaggi* (1944) and *The Injured Miner* (1963), which highlight the difficult and dangerous working conditions endured by miners.

George Bissell, *Coal Miners at Work, Cutting Coal and Propping*, between 1939 and 1946.

There are, however, also moments of humour in coal art. In 1921 the owners of Atherton Collieries in Lancashire asked artist W. Heath Robinson to record his impressions of their mining operations for a 1922 calendar. *The 'First' Colliery, An Almanac of Four Drawings*, which comprises 'A Busy Day in the Washery', 'Screening and Picking', 'The Pit' and 'Pit Head', as well as 'Just Up – An Old-time Miner Leaving Work' which adorns the cover, humorously transforms the mine sites he visited in typical Heath Robinson fashion. The cross-section of 'The Pit', shaped to resemble a one-pound sign (£1), may well be a wry comment on the money the mine owners are making at the expense of their miners' hard labour.

There is also a substantial body of illustrations in newspapers and magazines from the late nineteenth century that depicts the coal industry both above and below ground, as a brief snapshot reveals: in 1871 *The Graphic* carried M. W. Ridley's 'Pitmen Hewing Coal'; Francis S. Walker's 'Group of "Tip" Girls' featured in the *Illustrated London News* in 1875; W. H. Overend's 'Colliery Disaster at Seaham' appeared in the same periodical in 1880; in 1878 four illustrations by J. Nash, collectively titled 'Work at a Coal Mine, 1' were published in *The Graphic*; and A. S. Hartwick's engravings of miners at work featured in *The Daily Graphic* on two consecutive days in 1892. The Wigan pit brow girls, who feature prominently in Martin Cruz Smith's

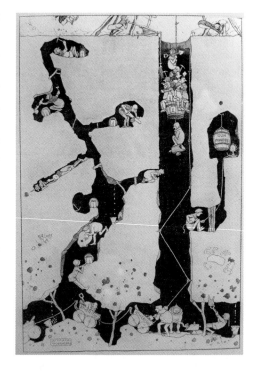

W. Heath Robinson, 'The Pit', sketch for *The 'First' Colliery: An Almanac of Four Drawings*, 1922 calendar for Atherton Collieries, Lancashire.

1996 thriller *Rose*, attracted the attention of periodical editors, with an 1874 cover of the *Pictorial World* given over to an image of 'Wigan Collieries: Women Working at the Coal Shoots'; and the *Illustrated London News* carried an unsigned illustration of 'Lancashire Pit-brow Women' in 1887. Female coal mine workers were also the subject of paintings and drawings, among them Hannah Keen's *Portrait of a Pit Brow Girl* (1895); several works by Meunier, including *Three Female Miners* (1885) and *A Female Miner* (1887); and works by Cécile Douard, notably *La hiercheuse* (1896).

In another medium, postcards of Wigan pit brow girls were produced by several photographic studios in the town in the second half of the nineteenth century; Arthur Munby collected hundreds of these photographs, which are now housed in the Munby Archives in the Trinity College Library at the University of Cambridge. While postcards carrying images of pit brow girls provide social testimony, photographic postcards of coal mines, coal miners and coal merchants going about their daily toil, coal ships, coal wharves and imperial British and American coaling stations register the commercial contexts of the industry. More broadly, the lives of coal miners and their families, from the northeast of England and South Wales to the Appalachians, have been eloquently and candidly captured by diverse photographers, including Bill Brandt, Bruce Davidson and Robert Frank. In a more overtly political context, Lewis Hine's photographs of child miners were famously used in the campaign for reform of the child labour laws in the United States in the early twentieth century.

W. H. Overend, 'The Colliery Disaster at Seaham: Explorers Descending the Pit to Rescue the Men Below', cover of the *Illustrated London News*, 18 September 1880.

Coal has also been a frequent subject of poster art in the twentieth century. As a key resource, coal was the subject of numerous propaganda posters during both world wars – urging miners to dig more coal or consumers to use less coal, to support the war effort – including Norman Rockwell's 1943 portrait of a coal miner. A 1936 poster of the head and shoulders of a coal miner by Isadore Posoff was produced for the Work Projects Administration Federal Art Project in Pennsylvania, where it was used to promote local tourism. Another prime example of coal poster art comes from the brush of English artist George Clausen. In 1924 the London, Midland & Scottish Railway Company (LMS) commissioned a series of sixteen posters by members of the Royal Academy, including Clausen's *Coal*, one of three in the series depicting 'British Industries'. In the foreground of the poster a line of miners leave a coal mine at the end of their shift. The colliery buildings and pithead winding gear spread across the background, while in the middle ground, behind the miners,

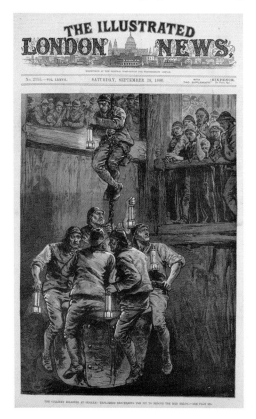

Norman Rockwell,
'Mine America's Coal',
poster produced by the
U.S. War Manpower
Commission, 1943.

Cécile Douard,
La hiercheuse
(Female Coal Miner),
1896, oil on canvas.

parallel lines of laden LMS coal wagons stretch out from the
colliery, demonstrating how the LMS serves British industry.

Beyond artists taking coal as their subject, miners took to
art, too. In the 1930s and '40s a growing number of miners –
who in the 1970s would retrospectively be collectively labelled
the 'Pitmen Painters' by art historian William Feaver – began
to paint what they knew from direct experience. As Robert
McManners and Gillian Wales observe in their book *Shafts of
Light: Mining Art in the Great Northern Coalfield*, while 'there

had been no great tradition in [Britain] of miners painting', that 'changed between the wars',[5] when miners seized the opportunities for adult learning afforded through groups such as the Workers Educational Association. McManners and Wales go on to explain:

> This new-found confidence resulted in an original and exciting body of experiential art created by the working classes – a movement of art which defined the culture and communities from which it had spawned. The miner, in particular, felt compelled to tell his story more graphically and comprehensively through the arts than the worker in any other industry.[6]

In discussing the work of miner-artists from the Yorkshire coalfields, Emily Green notes that 'works produced by coal miners portray the coal mining industry in many aspects, from the coalface through to the miners' home lives and also document the changing landscape of the English coalfields.'[7] This is true not only of the Yorkshire coalfield artists, but of all miner-artists, including the members of the Ashington Group and the artists of the socialist-inspired Spennymoor Settlement in County Durham.

Works by the Ashington Group, which began as an art appreciation class run by Durham University lecturer Robert Lyon, are now housed in a permanent gallery at Woodhorn Museum and Northumberland Archives in Ashington – a museum dedicated to coal mining. They include colliery landscapes, such as *Ashington Colliery*, a 1936 oil painting by founder member Harry Wilson, and William Crichton's *Pit Yard* (1947), while the headgear of a colliery stands in the middle background of George Blessed's well-known painting *Whippets* (c. 1939). Underground works include Leslie Brownrigg's *The Miner* (1935) and *Miner Setting Props in Lower Seam* (1950) by Oliver Kilbourn, another founder member and one of the last two surviving participants in the group, which met weekly from 1934 until it was finally disbanded in 1983.

Works by artists of the Spennymoor Settlement painting group feature prominently in the Gemini Art Collection of Mining Art, a collection of 420 paintings and drawings collected by McManners and Wales and now permanently housed in the Mining Art Gallery at Auckland Castle, the first museum gallery in Britain dedicated to coal art, which opened in 2017. The sheer physicality of working underground is captured in works like Robert Olley's *Off the Way* (2000) and Ted Holloway's *Setting Timber* (1983), while Holloway and Tom Lamb reflect on less muscular but potentially more dangerous moments underground in *Testing for Gas* from the 1950s and *The Shotfirer* from the 1990s, respectively. Above-ground scenes feature in numerous distinctive works by Norman Cornish, one of the most famous of the Pitmen Painters, including *Pithead* (n.d.), *Miners on the Pit Road* (n.d.) and *Pit Gantry* (n.d.), as well as *Study for Miners' Gala Mural* (1963), which reflects on the social activity in a mining community. Tom McGuinness, another celebrated

Robert Olley, *Off the Way*, 2000, oil painting.

member of the Pitmen Painters, offers a further comprehensive and striking record of the lives of miners in paintings such as *Miners in the Roadway* (1958) and *Salvage Men Brawing Rings* (1958), which depict work underground, as well as paintings like *Miners' Lamp Room* (1962), *Pithead Bath* (1958) and *Colliery Canteen* (n.d.) that capture moments of the miner's day above ground.

While the stories of Britain's Pitmen Painters have been widely disseminated, notably in William Feaver's 1988 book *Pitmen Painters: The Ashington Group, 1934–1984* and Lee Hall's play *The Pitmen Painters* (2007), the story of Japan's pitman painter, Sakubei Yamamoto, is less well known but equally fascinating. Yamamoto 'worked for more than half a century for almost 20 different mining companies'.[8] In more than five hundred coal paintings – now housed in the Tagawa City Coal Mining Historical Museum – he documented the story of Japanese coal miners of the Chikuho region from the late nineteenth century through to the middle of the twentieth century.

Sakubei Yamamoto, *Coal Mining in a Squatting Position* (no. 314).

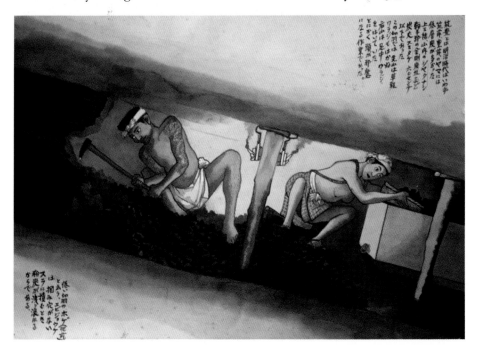

Hans Baluschek,
Coal Loads, 1901,
oil on canvas.

In 2011 his coal-mining paintings and writings were added to UNESCO's Memory of the World Register.

Valerie Ganz is one of relatively few female artists to have chosen the coal industry as a subject. A native of Swansea, Ganz was drawn to the industrial heritage of South Wales, and in particular the working lives of Welsh coal miners. Over several years in the 1980s she pursued her art in fourteen different collieries, including the Six Bells Colliery in Abertillery, where she spent a year sketching the miners at work both above ground and at the coalface, and another year working at three opencast mines on a commission for British Coal. In 1986 Ganz's work featured alongside that of renowned coal artists Josef Herman, Jack Crabtree and ex-miner Nicholas (Nick) Evans in 'Mining in Art', a major exhibition mounted at the Glynn Vivian Art Gallery in Swansea.[9] More recently, in Australia, artist Christine

Eddie McIlquham,
Bear no. 5, 2011,
Harlan County,
Kentucky.

Pike portrayed miners and mining in the Hunter Valley of New South Wales in her 2015 'Coal Miners' exhibition.

As well as the subject of art, coal can also be the medium of art. In Martin Booth's novel *The Industry of Souls* (1998), Avel, one of Alexander Bayliss's 'comrades in coal', 'passes his time carving chess pieces out of coal and shale'.[10] According to Jon McCaughey, 'Sculpting in coal, or coal carving as it was called, was all the rage in the north of England in the 17th and 18th centuries,'[11] though it has since all but died out as an art form. Today coal carving is rather a craft, producing souvenirs for sale in the gift shops of exhibition mines and elsewhere. But if coal carving is no longer a popular art form, the use of coal in art is thriving. At the 1938 'Exposition Internationale de Surréalisme' in Paris, Marcel Duchamp suspended 1,200 coal sacks from the gallery ceiling, from where they dropped coal dust on the exhibition's guests.[12] Greek-Italian artist Jannis Kounellis, a pioneer of the Arte Povera movement, regularly used coal or sacks of coal in his installations: examples include *Coal Sculpture with*

Jannis Kounellis,
Untitled, 1998, detail,
Museum of Old and
New Art, Berriedale,
Hobart, Australia.

Wall of Coloured Glass (1990–2005), where coal spills from a steel coal store in front of a coloured glass screen; *Untitled (Sack with Z)* (2001), in which a sack of coal with a large letter Z painted on the front is contained in a mesh-fronted metal box; and *Untitled* (1998), in which coils of rope (or sides of beef) hang on hooks above full sacks of coal at David Walsh's Museum of Old and New Art (MONA) in Berriedale, Tasmania.

In 2011 a public art project in Harlan County saw six painted bears placed around the county, each representing an aspect of life in the southeastern Kentucky community. Bear #5 is decorated with a coal theme by local artist and resident of Lynch, Eddie McIlquham. Encrusted with small pieces of coal, except for a bare patch on its side where three underground scenes capture some of the history of mining in the area, McIlquham's bear is a reminder that coal can be both the subject *and* the medium of art, and that while coal art is universal, like the coal industry it is also strongly rooted in place and community.

9 Coal and Folklore

Perhaps the best-known and most widespread cultural tradition involving coal is the practice of 'first footing'. I can still recall the excitement I felt as a child when I was first allowed to stay up until midnight on New Year's Eve, and the confusion I experienced when my father went outside just before midnight to return after the hour had struck carrying a piece of coal, which he proceeded to add to the fire. I learnt later that traditionally coal had been brought into homes in the north of England (preferably by a dark-haired man, as my father was then) as a sign of warmth and prosperity for the new year. First footing is also common across Scotland, where the first foot might also bring other symbolic gifts such as shortbread, salt, a black bun and (no surprise here) a dram of whisky!

In many Western cultures, children hang a Christmas stocking before going to bed on Christmas Eve, hoping that Father Christmas (or Santa Claus or St Nicholas) will visit during the night and that they will wake in the morning to find it filled with small gifts. In some of those cultures children are warned that if they behave badly during the year they will wake on Christmas morning to find that Father Christmas has left them a lump of coal in their stocking instead of the anticipated bounty of gifts. The origins of this belief are not clear. In the punishment version, after coming down the chimney Father Christmas discovers a naughty child lives in the house, and not knowing what to give him or her seizes on the nearest thing to hand – a piece of coal from the fireplace. Alternatively, coal is said to have been left for the children of the poor so they could warm themselves on cold winter nights, making it a welcome gift rather than a punishment. While the origin of the tradition is unknown, some suggest it is linked to Charles Dickens's *A Christmas Carol* (1843), which begins on Christmas Eve, with Ebenezer Scrooge working in his counting-house, and his clerk, Bob Cratchit, trying to warm himself over a candle as his employer refuses to give him more coal:

> The door of Scrooge's counting-house was open that he might keep his eye upon his clerk, who in a dismal little cell beyond, a sort of tank, was copying letters. Scrooge had a very small fire, but the clerk's fire was so very much smaller that it looked like one coal. But he couldn't replenish it, for Scrooge kept the coal-box in his own room; and so surely as the clerk came in with the shovel, the master predicted that it would be necessary for them to part. Wherefore the clerk put on his white comforter, and tried to warm himself at the candle; in which effort, not being a man of a strong imagination, he failed.[1]

Large piece of anthracite collected from a spoil heap in the village of Cwmllynfell, South Wales, in 1981.

In recent years the tradition has taken a new twist, with companies in both the United States and Australia offering to send gift-wrapped coal to a recipient of one's choice.[2]

All these traditions are linked to the association between coal and fire or warmth. Thus coal was also a symbolic gift given to newly married couples to bring them good fortune and also to newly installed occupants of a house to protect it from evil spirits. There are numerous other superstitions linked to coal and the home. The belief that to put shoes on a table is to tempt fate may have its origins in the tradition in mining communities in the north of England, where, when a miner died in a colliery accident, his shoes were put on the table as a mark of respect.

Ancient folk remedies for heartburn, indigestion or acid reflux included sucking a lump of coal.[3] Perhaps surprisingly, there is 'a lot of superstition in the theatre surrounding coal'. Alex Matsuo gives one striking example: 'If you stood onstage and threw coal in the back of the theatre and the balconies (typically known as the gallery), it would bring good luck for the production.'[4] And in Turkey coal and agriculture come together in several superstitions, including putting a small piece of coal into milk when giving it to somebody, otherwise the cow will stop producing milk, and mixing coal with the first milk after a cow has given birth to ensure its milk does not dry up.[5]

Coal has also been seen as a symbol of luck, and in the past (perhaps still?) many people carried a small piece of coal as a lucky charm. In *The Encyclopedia of Superstitions*, Richard Webster observes that in the nineteenth century burglars carried coal in the belief that it would prevent them from being caught, while sailors held that keeping a piece of coal in their

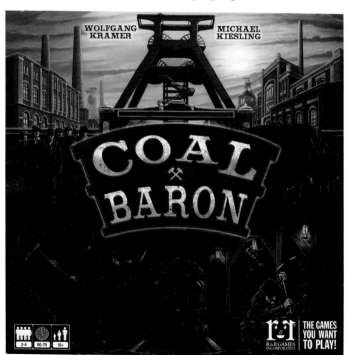

Coal Baron, a board-game designed by Michael Kiesling and Wolfgang Kramer, 2013.

pocket while at sea (preferably a piece found on a beach) would protect them from drowning.[6] Similarly, there is a tradition of soldiers carrying small lumps of coal into battle to ensure their survival. Owen Davies records that the Imperial War Museum in London 'possesses a lucky lump of coal carried by a soldier of the City of London Yeomanry. It had been sent to him by his sister in 1917.' He goes on to explain:

> This tradition, which had become popular by the early twentieth century, derives from the growth of coal as the staple hearth fuel of the industrial poor. Philip Gibbs wrote how, just prior to leaving for Ypres, an Irish officer took a little lump of coal out of his tunic and said in earnest, 'It will help you as it helped me. It's my lucky charm.'[7]

Indeed, Steve Roud observes that coal was widely regarded as a lucky object for much of the twentieth century, noting that the great folklorists Iona and Peter Opie discovered children taking coal into exams and adults carrying a piece during driving tests. Fishermen, too, particularly older fishermen, were careful to carry a piece of coal in their boats.[8] In Britain and Europe, chimney sweeps, with their coal-blackened faces, were believed to carry luck with them – and meeting a chimney sweep could result in some of his luck rubbing off on you, particularly on your wedding day. George L. Phillips notes that, 'not by mere chance was a sooty chimney-sweeper sauntering in front of Kensington Palace on the wedding morning of Prince Philip and Princess Elizabeth, thereby affording the excited bridegroom an opportunity to dash out from the royal apartments to wring his grubby hand for "sweep's luck".'[9]

However, as the domestic use of coal has declined, meaning children are less familiar with it, it is unlikely that many of these folkloric traditions linked to coal will survive far into the future. Similarly, while traditionally coal was used to provide the eyes (and a carrot the nose) of a snowman, other rocks and pebbles are already taking the place of the black stuff, which is no longer accessible to most children.

Advertisement for
Adams' Patent Coal
Sifter, 1859.

ADAMS' PATENT COAL SIFTER.

Thousands of tons of good Coal are wasted and thrown away annually, even by *economists*, rather than endure the annoyances of DUST, consequent upon the use of the ordinary Coal Sifter. This Patent Sifter overcomes all objections, and effects A SAVING OF AT LEAST ONE-FIFTH.

1st.—It can be used on a common Flour Barrel, or Iron Ash Barrel of same size.
2d.—The operation of sifting being performed under the cover, the dust is all confined.
3d.—The above cut represents a lady in her daily attire, sifting with perfect ease, and without injury to her clothing.
4th.—It is substantially made, and with tolerable care will endure for years.
5th.—It is warranted to give satisfaction, or money refunded, if returned within reasonable time.

PRICES:
SIFTER No. 2, FLOUR BARREL SIZE, - - - - - - $2.25
No. 2, INCLUDING TUB, or No. 4, INCLUDING KETTLE, $2.75
No. 2, INCLUDING IRON SAFE OR ASH BARREL, - - $3.50
DELIVERED IN ANY PART OF THE CITY FREE OF EXPENSE.

Orders addressed to

SANFORD ADAMS, 66 Cortlandt Street and Lovejoy's Hotel, N.Y.

They can also be obtained at the following places:
STEVENS BRO. & CO. 222 Pearl St.; J. & C. BERRIAN, 601 Broadway; REA & POLLOCK, 48 Cortlandt Street.

At the end of the seventeenth century, John Aubrey records in his *Miscellanies* (1696) some of the various superstitions young women believed would enable them to dream of the man who would be their future husband, including putting a piece of coal under their head at night.[10] Still on the subject of marriage superstitions, a live piece of coal falling out of the fire and landing by your feet was believed to be an indication that you would soon be getting married,[11] while in China some people believed that a bride stepping across a basin of burning coal when she

Postcard: Coal
Heritage Trail,
West Virginia –
Montage Scenes.

arrived at the home of the groom's family would burn away any
bad luck she may be carrying. Another household superstition
is that 'when a gaseous piece of coal makes a sputtering noise
it is a sign of a row; you should give it a vigorous stir.'[12] And
a piece of burning coal falling from the fireplace is a sign that
a stranger is coming.

In a talk on Indian superstitions presented to the Indian section of the Royal Society of Arts, 9 February 1911, R. A. Leslie Moore notes that among Goan Christians it is a good omen to see a piece of coal.[13] Other Indian superstitions include hanging a totem (string) of lemon and chilli (*nimbu-mirchi totka*), tied off with a piece of coal, outside homes and shops, or in vehicles, to ward off the evil eye.[14]

Beyond the folklore associated with coal, there is also folkloric tradition associated with coal miners and coal mines. Folklore studies record that since the Middle Ages there have been tales about mythical creatures and spirits inhabiting coal mines, particularly in British and German folklore. In British folklore these creatures were known as Knockers, or Bwcas in Wales. As James C. Baker explains, 'In the oral traditions from the coal-pits near Birmingham and Manchester, Knockers are benign little beings who work for the miner at night. They also warn the miner with tapping noises just before a rock fall.'[15] However, like most sprites or goblins of folklore, they could also be malevolent or mischievous. Their name was derived from the knocking sound (the creaking of earth and timbers) that often precedes a rock fall – either because they are good spirits trying to warn the miners of imminent danger, or because they are evil spirits hammering at walls and supports to cause the collapse. And some miners chose to blame missing or broken tools and extinguished lamps on the Knockers. The folklore was carried to the United States, probably by Cornish miners, where the little creatures are called Tommyknockers. In German folklore the Kobold, which could be heard working in distant parts of mines, is an equivalent of the Knocker.[16] In *Household Tales and Traditional Remains* (1895), Sidney Oldall Addy notes that 'colliers in the north of Derbyshire leave a hundredweight of coal in the pit every week for the fairies.'[17]

Apart from a belief in supernatural beings, which, of course, was much more prevalent in earlier centuries, miners had many other superstitions, too. In eastern Pennsylvania, for example, it was considered bad luck to enter a mine if you met a red-headed woman on the way to work. And in many mining regions of the

United States it was considered bad luck for a woman to enter a mine – as a number of miners point out in Ben Ketai's horror film *Beneath*, when retiring coal miner George brings his daughter Sam with him on his final shift in the mine. And indeed, since 1973, when women were allowed to work as miners in the United States, they have had to battle this superstition. In some coal regions, including Kentucky, it was believed that whistling in a coal mine brought bad luck, while it was good luck for a miner to get coal dust in his eyes.[18] Another belief among miners (with which visitors to the No. 9 Mine in Pennsylvania are regaled) was that the rats in the mines could sense an impending collapse or gas explosion, so that when they saw the rats running for safety, the miners ran, too. Rather more curiously, Addy records that in the Derbyshire coalfields, 'Colliers say that accidents are most frequent in coal-pits when broad beans are in bloom.'[19]

Many of the cultural traditions associated with coal reflect the importance it once held in the domestic sphere. Conversely, the superstitions of those whose lot has been to work the black seams over the centuries tend to reflect on the terrible price of coal.

Coal, Rhondda Heritage Park, South Wales.

Afterword

There is considerable irony in the fact that coal, the fuel with which we have had a passionate affair for over two centuries, now poses a grave danger to our future.

Shortly before Charles Smithson and Sarah Woodruff consummate their relationship in John Fowles's *The French Lieutenant's Woman* (1969), 'a little cascade of coals' falls from the fire and sets the blanket covering Sarah's legs smouldering.[1] At this moment in the novel, coal operates as a metaphor both for their passion and for the danger inherent in that passion. Fowles's neo-Victorian, postmodernist historical novel is famous for the three endings the author presents to his readers. The first ending is the marriage of Charles and Ernestina, a 'thoroughly traditional ending', which Fowles promptly insists is 'not what happened'.[2] In the second ending, Charles and Sarah are reunited after two years apart. And in the third ending (which concludes the novel), Charles again finds Sarah after a two-year search, but this time she rejects him. Ultimately, Fowles leaves the reader to choose whichever ending, if any, he or she wishes; unlike Victorian novels, postmodern novels do not require closure.

Like *The French Lieutenant's Woman*, the story of coal has multiple possible endings. A thoroughly conventional ending is that we immediately end our affair with coal. But despite news that in May 2019 Britain went a week without coal power for the first time since 1882,[3] most readers of this book will agree that this ending can be set aside as readily as Fowles sets aside his happy Victorian closure. A second, darker ending, presaged, for example, by Australia's determination to approve and build new mines in the Galilee Basin and elsewhere, is that, after a hiatus, we will continue our affair with coal. A third ending is that we continue what we know to be a sordid affair with coal for a while longer, before breaking it off for good.

The conclusion, dear readers, is of our making.

Lancashire Mining Museum, Astley Green Colliery: the only surviving pit headgear on the Lancashire coalfield.

REFERENCES

Preface

1 Alexander Watt, *The History of a Lump of Coal: From the Pit's Mouth to a Bonnet Ribbon* (London, 1882), p. 1.
2 Neil Gunson, ed., *Australian Reminiscences and Papers of L. E. Threlkeld, Missionary to the Aborigines, 1824–1859* (Canberra, 1974), p. 64.
3 See Barbara Freese, *Coal: A Human History*, 2nd edn (New York, 2016), pp. 27–8.
4 Ibid., p. 28.
5 Jules Verne, *The Child of the Cavern; or, Strange Doings Underground* [1877] (London, n.d.), p. 103.
6 David Jones, 'Coal Furniture in Scotland', *Furniture History*, XXIII (1987), p. 36.

1 What Is Coal?

1 Jules Verne, *The Mysterious Island* (New York, 1920), p. 251.
2 'Mines and Minerals Law and Legal Definition', *usLegal*, www.uslegal.com, accessed 15 January 2019.
3 James M. Schopf, 'A Definition of Coal', *Economic Geology*, LI/6 (1956), p. 527.
4 Jack A. Simon and M. E. Hopkins, 'Geology of Coal', in *Elements of Practical Coal Mining*, ed. Samuel M. Cassidy (New York, 1973), p. 11.
5 'Coal Geology', *Geoscience Australia*, www.ga.gov.au, accessed 14 January 2019.
6 H. Stanley Jevons, *The British Coal Trade* (London, 1915), p. 31.
7 See W. G. Chaloner, 'The Palaeobotanical Work of Marie Stopes', in *History of Palaeobotany: Selected Essays*, ed. A. J. Bowden, A. J. Burek and R. Wilding (London, 2005), pp. 127–35.
8 'Macerals', *Kentucky Geological Survey*, www.uky.edu, accessed 14 January 2019.

9 Jennifer M. K. O'Keefe et al., 'On the Fundamental Difference between Coal Rank and Coal Type', *International Journal of Coal Geology*, CXVIII (2013), p. 59.
10 'Coal', *Encyclopaedia Britannica*, www.britannica.com, accessed 15 January 2019.
11 Kate Grenville, 'A Short History of Coal', www.kategrenville.com. au, accessed 23 July 2018.
12 O'Keefe et al., 'Coal Rank and Coal Type', p. 74.
13 Alexander Watt, *The History of a Lump of Coal: From the Pit's Mouth to a Bonnet Ribbon* (London, 1882), p. 60.
14 Ibid., pp. 60–61.
15 Heike Egner, Marén Schorch and Martin Voss, *Learning and Calamities: Practices, Interpretations, Patterns* (London, 2015), p. 107.
16 C. William Siemens, FRS, 'Science in Relation to the Arts', *Popular Science Monthly*, XXII (1882), p. 209.
17 Egner, Schorch and Voss, *Learning and Calamities*, p. 102.
18 Ibid., pp. 103–4.
19 Ibid., p. 107.
20 George Orwell, 'The Case for the Open Fire', *Evening Standard* (8 December 1945), p. 6.
21 G. S. Callendar, 'The Artificial Production of Carbon Dioxide and Its Influence on Temperature', *Quarterly Journal of the Royal Meteorological Society*, LXIV/275 (1938), pp. 223–40.
22 Karen Pinkus, *Fuel: A Speculative Dictionary* (Minneapolis, MN, 2016), p. 48.
23 Barbara Freese, *Coal: A Human History*, 2nd edn (New York, 2016), p. 7.

2 Using Coal

1 Peter J. Golas, *Science and Civilisation in China*, vol. V: *Chemistry and Chemical Technology, Part 13, Mining*, ed. Joseph Needham (Cambridge, 1999), p. 190.
2 Ibid., pp. 191–4.
3 Ibid., p. 195.
4 Marco Polo, *The Travels of Marco Polo the Venetian* (London, 1908), p. 215.
5 See Samuel M. Cassidy, 'History of Coal Mining', in *Elements of Practical Coal Mining*, ed. Samuel M. Cassidy (New York, 1973), p. 1.
6 Robert Galloway, *Annals of Coal Mining and the Coal Trade*, vol. I (London, 1898), pp. 6–7.
7 Barbara Freese, *Coal: A Human History*, 2nd edn (New York, 2016), pp. 15–16.

8 Galloway, *Annals of Coal Mining*, pp. 5–7.

9 Bede the Venerable, *The Venerable Bede's Ecclesiastical History of England, and also the Anglo-Saxon Chronicles*, ed. J. A. Giles (London, 1849), p. 5.

10 Galloway, *Annals of Coal Mining*, pp. 17–18.

11 John Hatcher, *The History of the British Coal Industry*, vol. 1: *Before 1700: Towards the Age of Coal* (Oxford, 1993), pp. 24–5; Peter Brimblecombe, *The Big Smoke: A History of Air Pollution in London since Medieval Times* (London, 1987), p. 7; Galloway, *Annals of Coal Mining*, pp. 29–30.

12 J. U. Nef, *The Rise of the British Coal Industry* (London, 1932), p. 201.

13 John Milton, 'At a Vacation Exercise in the College, Part Latin, Part English', in *The Major Works*, ed. Stephen Orgel and Jonathan Goldberg (Oxford, 2003), p. 78.

14 Andrew C. F. David, 'Cook, James', *Oxford Dictionary of National Biography*, www.oxforddnb.com, accessed 24 January 2019.

15 Freese, *Coal*, p. 25.

16 Galloway, *Annals of Coal Mining*, pp. 46, 54, 63–4.

17 Nef, *Rise of the British Coal Industry*, p. 201.

18 W. G. [William Gray], *Chorographia; or, A Survey of Newcastle-upon-Tyne: 1649* (Newcastle upon Tyne, 1884), pp. 90–91.

19 Michael Ondaatje, *Warlight* (London, 2018), p. 33.

20 Joel Mokyr, *The Enlightened Economy: An Economic History of Britain, 1700–1850* (New Haven, CT, 2009), p. 22.

21 Martin Ele, 'An Account of the Making Pitch, Tar, and Oil Out of a Blackish Stone in Shropshire, Communicated by Mr. Martin Ele the Inventor of It', *Philosophical Transactions of the Royal Society*, XIX/228 (1695), p. 544, https://royalsocietypublishing.org, accessed 31 January 2019.

22 See Christine L. Corton, *London Fog: The Biography* (Cambridge, MA, 2015), p. 2.

23 Mokyr, *The Enlightened Economy*, p. 100; Freese, *Coal*, p. 66.

24 Freese, *Coal*, pp. 43–4.

25 Heidi C. M. Scott, *Fuel: An Ecocritical History* (London, 2018), p. 121.

26 Roy Church, with Alan Hall and John Kanefsky, *The History of the British Coal Industry*, vol. III: *1830–1913: Victorian Pre-eminence* (Oxford, 1986), p. 91.

27 George Orwell, *The Road to Wigan Pier* [1937] (London, 1988), p. 19.

28 G. K. Chesterton, *Tremendous Trifles* (London, 1909), p. 158.

29 Freese, *Coal*, p. 73.

30 See John K. Walton, *Lancashire: A Social History, 1558–1939* (Manchester, 1987).

31 Freese, *Coal*, p. 107.

32 See ibid., pp. 108–10.

33 See Christopher F. Jones, *Routes of Power: Energy and Modern America* (Cambridge, MA, 2014), pp. 23–58.

34 See Priscilla Long, *Where the Sun Never Shines: A History of America's Bloody Coal Industry* (New York, 1989), p. xxiii.

35 Alexander Watt, *The History of a Lump of Coal: From the Pit's Mouth to a Bonnet Ribbon* (London, 1882), p. 83.

36 Mokyr, *The Enlightened Economy*, p. 136.

37 Paula E. Dumas, *Proslavery Britain: Fighting for Slavery in an Era of Abolition* (New York, 2016), p. 23.

38 See Stephen Gray, *Steam Power and Sea Power: Coal, the Royal Navy, and the British Empire, c. 1870–1914* (London, 2018), pp. 1–6.

39 Anne McClintock, *Imperial Leather: Race, Gender and Sexuality in the Colonial Contest* (New York and London, 1995), p. 33.

40 Jules Verne, *The Child of the Cavern; or, Strange Doings Underground* (London, n.d.), p. 10.

41 See statistical data set, 'Historical Coal Data: Coal Production, Availability and Consumption'. www.gov.uk, accessed 11 February 2020.

42 Orwell, *The Road to Wigan Pier*, p. 29.

43 'U.S. Coal Consumption in 2018 Expected to Be the Lowest in 39 Years', www.eia.gov, 4 December 2018.

44 See 'Coal & Electricity', www.worldcoal.org, accessed 13 January 2020.

45 See 'Electricity Generation', www.energy.gov.au, accessed 13 January 2020.

46 See 'Coal Resources', www.energy.gov.za, accessed 13 January 2020.

47 See 'Coal's Share of China Electricity Generation Dropped below 60% in 2018', https://ieefa.org, accessed 13 January 2020.

48 See 'Uses of Coal', www.worldcoal.org, accessed 6 February 2019.

49 H. Stanley Jevons, *The British Coal Trade* (London, 1915), pp. 31–2.

50 See Jiayu Wang, 'How Green Are Electric Vehicles?', https://sydney.edu.au, 6 May 2019.

3 Working the Black Seam

1 D. H. Lawrence, *Sons and Lovers*, ed. David Trotter (Oxford, 2009), p. 5.

2 Jules Verne, *The Mysterious Island* [1874] (New York, 1920), p. 94.

3 J.C., *The Compleat Collier; or, The Whole Art of Sinking, Getting, and Working, Coal-mines, &c. As is now used in the Northern Parts, Especially about Sunderland and New-Castle* (London, 1708), pp. 42–3.

4 See also John Hatcher, *The History of the British Coal Industry*, vol. 1: *Before 1700: Towards the Age of Coal* (Oxford, 1993), p. 205.

5   Ibid., pp. 206–7.
6   Ibid., p. 207.
7   See J.C., *The Compleat Collier*, pp. 34–7.
8   Clare Garner, 'Last Pit Ponies Are Made Redundant',
    The Independent (15 February 1999), p. 8.
9   See Samuel M. Cassidy, 'History of Coal Mining', in *Elements of
    Practical Coal Mining*, ed. Samuel M. Cassidy (New York, 1973), p. 2.
10 Karen Pinkus, *Fuel: A Speculative Dictionary* (Minneapolis, MN,
    2016), p. 49.
11 Robert Galloway, *Annals of Coal Mining and the Coal Trade*, vol. I
    (London, 1898), p. 160.
12 Roy Church, with Alan Hall and John Kanefsky, *The History of
    the British Coal Industry*, vol. III: *1830–1913: Victorian Pre-eminence*
    (Oxford, 1986), p. 322.
13 Frances Hodgson Burnett, *That Lass o' Lowrie's* (New York, 1895),
    p. 53.
14 See www.group.rwe, accessed 5 March 2019.
15 See 'Coal', British Geological Survey, March 2010, www.bgs.ac.uk,
    accessed 4 March 2019; *Underground Coal*, www.undergroundcoal.
    com.au, accessed 4 March 2019; *Kentucky Coal Education*, www.
    coaleducation.org, accessed 4 March 2019.
16 Arthur McIvor and Ronald Johnston, *Miners' Lung: A History of
    Dust Disease in British Coal Mining* (Aldershot, 2007), p. 35.
17 Ibid.
18 John Worthen, *D.H. Lawrence: The Early Years, 1885–1930*
    (Cambridge, 1991), p. 11.
19 Upton Sinclair, *King Coal* [1917] (n.p., 2007), p. 26.
20 William Stewart, *J. Keir Hardie: A Biography* (London, 1921), p. 6.
21 Elizabeth Barrett Browning, 'The Cry of the Children', *Blackwood's
    Edinburgh Magazine*, LIV/334 (August 1843), p. 261.
22 Peter Kirby, *Child Labour in Britain, 1750–1870* (Basingstoke, 2003),
    p. 110.
23 See Rina Chandran, 'Children Working in India's Coal Mines
    Came as a "Complete Shock", Filmmaker Says', www.reuters.com,
    accessed 6 March 2019.
24 See Yaqub Azorda, 'Child Labour in Afghan Coal Mines', https://
    iwpr.net, accessed 7 March 2019; Michelle Nichols, 'Afghanistan
    Vows to "Set Standards" on Child Labor in Mines', www.reuters.
    com, accessed 7 March 2019.
25 John Hannavy, 'Amazons among the Coal Tubs', *History Today*,
    CIV/1 (2004), p. 27.
26 Ibid., p. 28.
27 See ibid., pp. 27–9; Alan Davies, *The Pit Brow Women of Wigan
    Coalfield* (Stroud, 2006); Dave Lane, *Pit Brow Lasses* (n.p., 2007);

Carletta Savage, 'Re-gendering Coal: Female Miners and Male Supervisors', *Appalachian Journal*, xxvii/3 (2000), p. 233; Isabel Moussalli, 'Women Were Only Let into Underground Mines 30 Years Ago, So Why Are They Leaving the Industry?', www.abc.net. au, accessed 14 March 2019.

28 See Savage, 'Re-gendering Coal', pp. 232–5.

29 Kuntala Lahiri-Dutt, 'The Shifting Gender of Coal: Feminist Musings on Women's Work in Indian Collieries', *South Asia: Journal of South Asian Studies*, xxxv/2 (2012), p. 462.

30 William B. Thesing and Ted Wojtasik, 'Poetry, Politics, and Coal Mines in Victorian England: Elizabeth Barrett Browning, Joseph Skipsey, and Thomas Llewelyn Thomas', in *Caverns of Night: Coal Mines in Art, Literature, and Film*, ed. William B. Thesing (Columbia, sc, 2000), p. 38.

31 McIvor and Johnston, *Miners' Lung*, p. 35.

32 Richard Llewellyn, *How Green Was My Valley* [1939] (London, 2001), p. 48.

33 Ibid.

34 Laurie Lee, 'The Village that Lost Its Children', in *I Can't Stay Long* (London, 1975), p. 86.

35 Quoted in Ceri Jackson, 'Aberfan: The Mistake that Cost a Village Its Children', www.bbc.co.uk, accessed 28 February 2019.

36 Lee, 'The Village that Lost Its Children', p. 98.

37 'Nord-Pas de Calais Mining Basin', whc.unesco.org, accessed 4 March 2019.

38 George Orwell, *The Road to Wigan Pier* [1937] (London, 1988), p. 94.

4 The Politics of Coal

1 Commonwealth of Australia, Parliamentary Debates, House of Representatives Official Hansard, 9 February 2019, p. 536, https:// parlinfo.aph.gov.au, accessed 1 April 2019.

2 Émile Zola, *Germinal*, trans. Peter Collier, intro. Robert Lethbridge (Oxford, 2008), p. 282.

3 Barbara Freese, *Coal: A Human History*, 2nd edn (New York, 2016), pp. 45–6.

4 Stephen Knight, 'Black Diamonds and Dust', *The Age* (Melbourne) (3 September 2005), p. 4.

5 A. J. Taylor, 'The Miners' Association of Great Britain and Ireland, 1842–48: A Study in the Problem of Integration', *Economica*, xxii/85 (1955), p. 46.

6 Ibid., pp. 59–60.

7 David Smith, 'Tonypandy 1910: Definitions of Community', *Past and Present*, lxxxvii/1 (1980), p. 158.

8 Ibid., p. 159.

9 Gwyn Williams, *When Was Wales?* (Harmondsworth, 1985), p. 222.

10 See Jim Phillips, 'The 1972 Miners' Strike: Popular Agency and Industrial Politics in Britain', *Contemporary British History*, xx/2 (2006), pp. 187–207.

11 See Hywel Francis and David Smith, *The Fed: A History of the South Wales Miners in the Twentieth Century* (London, 1980).

12 See Peter Gibbon, 'Analysing the British Miners' Strike of 1984–5', *Economy and Society*, xvii/2 (1988), p. 148.

13 Brian McCabe, 'Coal', in *Body Parts* (Edinburgh, 1999), p. 26.

14 J. E. George, 'The Coal Miners' Strike of 1897', *Quarterly Journal of Economics*, xii/2 (1898), pp. 187–8.

15 See Elliott J. Gorn, *Mother Jones: The Most Dangerous Woman in America* (New York, 2001); Rosemary Feurer, 'Mother Jones: A Global History of Struggle and Remembrance, from Cork to Illinois', *Illinois Heritage*, xvi/3 (2013), pp. 28–33.

16 See Lawrence R. Lynch, 'The West Virginia Coal Strike', *Political Science Quarterly*, xxix/4 (1914), p. 626–63.

17 Hoyt N. Wheeler, 'Mountaineer Mine Wars: An Analysis of the West Virginia Mine Wars of 1912–1913 and 1920–1921', *Business History Review*, l/1 (1976), p. 82.

18 Ibid., pp. 89–90.

19 Thomas G. Andrews, *Killing for Coal: America's Deadliest Labor War* (Cambridge, MA, 2008), p. 1.

20 Jessica Legnini, 'Radicals, Reunion, and Repatriation: Harlan County and the Constraints of History', *Register of the Kentucky Historical Society*, cvii/4 (2009), p. 494.

21 'Coal Resources', www.ga.gov.au, accessed 1 May 2019.

22 Cameron Amos and Tom Swann, 'Carmichael in Context: Quantifying Australia's Threat to Climate Action', www.tai.org.au, November 2015.

23 Dan Conifer, 'Adani Coal Mine a Step Closer with Environment Minister Endorsing Groundwater Approvals', www.abc.net.au, 9 April 2019.

24 Selina Ward, 'Dumping Abbot Point Dredge Spoil on Land Won't Save the Reef', https://theconversation.com, 17 March 2015.

25 See Michael Slezak, 'Clive Palmer Seeks Approval for "Monster Mine" Next Door to Adani', www.abc.net.au, 26 April 2018; Anne Davies, 'Clive Palmer's Coalmine Plan Scrutinised over Impact on Great Barrier Reef', www.theguardian.com, 22 May 2018; Mark Ludlow, 'Galilee Basin Coal Mines Will Go Ahead without Adani, Says Clive Palmer', www.afr.com, 1 March 2018.

26 Ewen Hosie, 'Clive Palmer Announces $1.5bn Coal-fired Power Station in QLD', www.australianmining.com.au, 11 September 2018.

27 See Tom Iggulden and Amy Greenbank, 'Plans for Chinese-backed Coal-fired Plant in NSW's Hunter Valley Could Reignite the Climate Wars', www.abc.net.au, 6 March 2019; Ben Smee, 'Deal Signed for Huge Coal-fired Power Plants in Hunter Valley, Hong Kong Firm Says', www.theguardian.com, 6 March 2019.

28 See www.stopadanialliance.com, accessed 2 May 2019.

29 Richard Flanagan, 'I'm Willing to Go to Jail to Stop Adani and Save Our Beloved Country. Will You Stand with Me?', www.theguardian.com, 5 May 2019.

30 'We Are the Limits! Anti-coal Protests across Czech Republic', www.foeeurope.org, accessed 6 May 2019.

31 See Oliver Milman, 'U.S. to Stage Its Largest Ever Climate Strike: "Somebody Must Sound the Alarm"', www.theguardian.com, 20 September 2019; Sabrina Barr, 'Climate Strike 2019: When Are the Global Protests and How Can You Take Part?', www.independent. co.uk, 27 September 2019; Matthew Taylor, Jonathan Watts and John Bartlett, 'Climate Crisis: 6 Million People Join Latest Wave of Global Protests', www.theguardian.com, 27 September 2019.

32 Adam Vaughan, 'Germany Agrees to End Reliance on Coal Station by 2038', www.theguardian.com, 26 January 2019.

33 See 'Coal Use Falls Again in Europe in 2017', http://www.eiu.com, accessed 6 May 2019.

34 See Avishek Rakshit, 'CIL Plans Expansion, Begins Talks with Bangladesh for Coal Exploration', www.business-standard.com, accessed 6 May 2019.

35 Richard Martin, *Coal Wars: The Future of Energy and the Fate of the Planet* (New York, 2015), p. 160.

36 Shellen Xiao Wu, *Empires of Coal: Fueling China's Entry into the Modern World Order, 1860–1920* (Stanford, CA, 2015), p. 198.

37 'Coal Explained: Coal and the Environment', www.eia.gov, accessed 7 May 2019.

38 Jasper Jolly, 'Britain Passes One Week without Coal Power for First Time since 1882', www.theguardian.com, 8 May 2019; Phoebe Weston, 'Britain Goes Week without Coal Power for First Time since Industrial Revolution', www.independent.co.uk, 8 May 2019.

39 Arthur Neslen, 'Spain to Close Most Coalmines in €250m Transition Deal', www.theguardian.com, 26 October 2018; 'Spain to Close Most Coal Mines by End of 2018', www.mining.com, 26 October 2018.

5 Coal Heritage and Tourism

1 Jules Verne, *The Child of the Cavern; or, Strange Doings Underground* [1877] (London, n.d.), p. 97.

2 See www.showcaves.com, accessed 25 June 2019.
3 Stephen Wanhill, 'Mines – A Tourist Attraction: Coal Mining in Industrial South Wales', *Journal of Travel Research*, XXXIX/1 (2000), p. 60.
4 'Princess Margaret Descends Pit', *The Times* (8 April 1954), p. 3.
5 Email from John Liffen, Curator Emeritus, Science Museum, London, 26 June 2019.
6 Jessica Legnini, 'Radicals, Reunion, and Repatriation: Harlan County and the Constraints of History', *Register of the Kentucky Historical Society*, CVII/4 (2009), p. 504.
7 Wanhill, 'Mines – A Tourist Attraction', p. 62.
8 'Major Mining Sites of Wallonia', whc.unesco.org, accessed 4 March 2019.
9 See 'Nord-Pas de Calais Mining Basin', whc.unesco.org, accessed 4 March 2019; Henry Samuel, 'France's Slag Heaps Join Pyramids on List of UNESCO World Treasures', www.telegraph.co.uk, 1 July 2012; Patrick Barkham, 'It's Just a Slag Heap, Isn't It? The World Heritage Sites that Defy Belief', www.theguardian.com, 2 July 2012.
10 Mining History Centre, www.chm-lewarde.com, accessed 1 July 2019.
11 Denise Cole, 'Exploring the Sustainability of Mining Heritage Tourism', *Journal of Sustainable Tourism*, XII/6 (2004), p. 484.
12 Ibid.
13 See 'Zollverein Coal Mine Industrial Complex in Essen', whc. unesco.org, accessed 26 June 2019.
14 Marcus Clarke, *For the Term of His Natural Life* [1874] (Melbourne, 2016), p. 336.
15 'Meet William Thompson', *Port Arthur Historic Sites*, www. portarthur.org.au, accessed 27 February 2019.
16 See https://nationalminingmuseum.com, accessed 25 June 2019.
17 'Coal Industry Centre – Tours', www.coalcentre.net, accessed 17 July 2019.

6 Coal in Literature

1 Wilfred Owen, 'Miners', in *The Collected Poems of Wilfred Owen* (London, 1974), p. 91.
2 Dylan Thomas, 'Reminiscences of Childhood', in *Quite Early One Morning: Poems, Stories, Essays* [1945] (London, 1987), pp. 1, 8; Dylan Thomas, *Under Milk Wood* [1954] (London, 1977), p. 1.
3 Jennifer Maiden, 'Coal', *The Age* (Melbourne) (28 August 2010).
4 Liz Berry, 'Homing', in *Black Country* (London, 2014), p. 4.
5 William Blake, 'The Chimney Sweeper', in *Selected Poetry*, ed. Michael Mason (Oxford, 2008), p. 69.

6 Sean O'Casey, *Juno and the Paycock*, in *Three Dublin Plays*, intro. Christopher Murray (London, 1998), pp. 88–9.

7 Charles Dickens, *Hard Times*, ed. Paul Schlicke (Oxford, 2006), p. 26.

8 A. J. Cronin, *The Stars Look Down* [1935] (London, 2013), pp. 10–11.

9 D. H. Lawrence, *Women in Love* [1920] (Harmondsworth, 1975), p. 12.

10 Upton Sinclair, *King Coal* [1917] (n.p., 2007), p. 142.

11 William Shakespeare, 'Venus and Adonis' (l. 533).

12 Thomas Gray, 'The Descent of Odin', in *Selected Poems*, ed. John Heath-Stubbs (Manchester, 1986), p. 66.

13 Alfred Lord Tennyson, *Selected Poems*, ed. Christopher Ricks (London, 2007), p. 11.

14 R. M. Ballantyne, *The Coral Island*, ed. Ralph Crane and Lisa Fletcher (Richmond, VA, 2015), pp. 141, 234.

15 Raymond Chandler, *The Big Sleep*, intro. Ian Rankin (London, 2005), p. 2.

16 Emily Brontë, *Wuthering Heights*, ed. Ian Jack, intro. Helen Small (Oxford, 2009) p. 61.

17 Émile Zola, *Germinal*, trans. Peter Collier, intro. Robert Lethbridge (Oxford, 2008), p. 40.

18 William Shakespeare, *Henry IV, Part 2* (II.i.63); Lord Byron, 'Beppo', in *The Major Works*, ed. Jerome J. McGann (Oxford, 2008), p. 328.

19 Charlotte Brontë, *Jane Eyre*, ed. Stevie Davis (London, 2006), p. 222; Martin Cruz Smith, *Rose* [1996] (London, 2014), p. 3.

20 Zola, *Germinal*, p. 76.

21 George Eliot, *The Mill on the Floss*, ed. Gordon S. Haight, intro. Juliette Atkinson (Oxford, 2015), p. 7.

22 James Lee Burke, *To the Bright and Shining Sun* [1971] (London, 2012), p. 6.

23 Denise Giardina, *Storming Heaven* [1987] (New York, 1988), pp. 166, 90.

24 Smith, *Rose*, p. 35.

25 Zola, *Germinal*, p. 307.

26 Jules Verne, *The Child of the Cavern; or, Strange Doings Underground* [1877] (London, n.d.), p. 9.

27 Ibid., p. 10.

28 G. A. Henty, *Facing Death or, the Hero of the Vaughan Pit: A Tale of the Coal Mines* [1882] (London, n.d.), p. 303.

29 Ibid., pp. 239–40.

30 Amy Lowell, 'The Coal Picker', *Poetry*, IV/5 (1914), p. 178.

31 Harold C. Wells, *The Earth Cries Out* (Sydney, 1950), p. 115.

32 Burke, *To the Bright and Shining Sun*, pp. 2, 10.

33 Richard Llewellyn, *How Green Was My Valley* [1939] (London, 2001), p. 200.
34 Smith, *Rose*, p. 21.
35 Zola, *Germinal*, p. 425.
36 Brian Nelson, 'Émile Zola (1840–1902): Naturalism', in *The Cambridge Companion to European Novelists*, ed. Michael Bell (Cambridge, 2012), p. 229.
37 Zola, *Germinal*, p. 15.
38 Ibid., pp. 32, 28, 27, 514.
39 Sinclair, *King Coal*, p. 114.
40 Llewellyn, *How Green Was My Valley*, pp. 146, 383.
41 Ken Follett, *Fall of Giants* [2010] (London, 2011), p. 17.
42 Smith, *Rose*, p. 35.
43 Ibid., p. 460.
44 Ibid., p. 31.
45 Ibid., pp. 31–2.
46 Angela V. John, *By the Sweat of Their Brows: Women Workers at Victorian Coal Mines* (London, 2006), p. 97.
47 See Smith, *Rose*, pp. 442–4.
48 Orpheus, 'A Pit Brow Wench for Me', *The Comet: A Fortnightly Journal of Fact, Fiction, and Free Opinion*, II (26 January 1889), p. 6.
49 Christina G. Rossetti, *Sing-song: A Nursery Rhyme Book*, with 120 illustrations by Arthur Hughes (London, 1893), p. 101.
50 Christina Rossetti, 'A Diamond or a Coal?', in *Complete Poems*, ed. Betty S. Flowers (London, 2001), p. 250.
51 Lorraine Janzen Kooistra, *Christina Rossetti and Illustration: A Publishing History* (Athens, OH, 2002), p. 113.

7 **Coal in Song and Film**

1 A. L. Lloyd, 'Introduction', in *Come All Ye Bold Miners: Ballads and Songs of the Coalfields*, compiled A. L. Lloyd (London, 1952), p. 11.
2 Rob Young, '"Jowel, Jowel and Listen Lad": Vernacular Song and the Industrial Archaeology of Coal Mining in Northern England', *Historical Archaeology*, XLVIII/1 (2014), p. 60.
3 Andy Irvine and Luke Plumb, sleeve notes, *Precious Heroes* (AK Records, 2016).
4 Andy Irvine, 'Hard Times in 'Comer's Mines', on *Precious Heroes*; lyrics reprinted with permission.
5 'The Blackleg Miners', in *Come All Ye Bold Miners*, ed. Lloyd, p. 99.
6 J. B. Geoghegan, 'Down in a Coal Mine' (1872), www.memory.loc. gov. pdf, accessed 9 March 2018.

7 Henri Storck and Joris Ivens, *Misère au Borinage* (1933); Patric Jean, *Les Enfants du Borinage, Lettre à Henri Storck* (1999), www.youtube. com, accessed 18 July 2019.
8 Stephen Beresford, *Pride*, shooting script, pp. 53–4, www. downloads.bbc.co.uk, accessed 11 July 2018.

8 Coal in Art

1 Douglas Gray, 'Art and Coal', in *Coal: British Mining in Art, 1680–1980*, exh. cat., organized by the Arts Council of Great Britain with the National Coal Board, City Museum and Art Gallery, Stoke-on-Trent, Glynn Vivian Art Gallery, Swansea, Science Museum London, D.L.I. Museum and Arts Centre, Durham, and Castle Museum, Nottingham (London, 1982), p. 10.
2 Mike McKiernan, 'Claude Monet *Les charbonniers* also called *Les déchargeurs de charbon* [The Coalmen, also called Men Unloading Coal] *c.* 1875', *Occupational Medicine*, LIX/2 (2009), p. 77.
3 'Claude Monet, *The Coalmen*', www.musee-orsay.fr, accessed 29 January 2018.
4 Linda Nochlin, *The Politics of Vision: Essays on Nineteenth-century Art and Society* (New York, 2018), p. 122.
5 Robert McManners and Gillian Wales, *Shafts of Light: Mining Art in the Great Northern Coalfield* (Durham, 2002), p. 18.
6 Ibid., p. 19.
7 Emily Green, 'Coal Mining and Art', in *The Hidden Artists of Barnsley* (Huddersfield, 2014), p. 282.
8 Diana Cooper-Richet, 'The "Pitmen Painters" of England and Japan', https://theconversation.com, 17 January 2018.
9 See www.valerieganz.co.uk, accessed 13 August 2018.
10 Martin Booth, *The Industry of Souls* (New York, 2000), pp. 30, 31.
11 Jon McCaughey, 'Coal Sculpture', *Design*, LXXII/4 (1971), p. 14.
12 Elena Filipovic, 'A Museum that Is Not', *e-flux Journal*, IV (2009), p. 5, www.e-flux.com, accessed 2 August 2018.

9 Coal and Folklore

1 Charles Dickens, *A Christmas Carol and Other Christmas Books*, ed. Robert Douglas-Fairhurst (Oxford, 2006), p. 11.
2 See, for example, www.sendcoal.org, accessed 18 March 2019.
3 See Max Cryer, *Superstitions: And Why We Have Them* (Auckland, 2016), p. 71.
4 Alex Matsuo, *The Haunted Actor: An Exploration of Supernatural Belief through Theatre* (Bloomington, IN, 2014), p. 135.

5 See 'Good Luck – Bad Luck', www.ktb.gov.tr, accessed 19 March 2019.
6 Richard Webster, *The Encyclopedia of Superstitions* (Woodbury, MN, 2008), p. 62.
7 Owen Davies, *A Supernatural War: Magic, Divination, and Faith during the First World War* (Oxford, 2018), p. 143.
8 Steve Roud, *The Penguin Guide to the Superstitions of Britain and Ireland* (London, 2003), pp. 103–4.
9 George L. Phillips, 'Sweep's Luck', *Notes and Queries*, CXCV (1950), p. 168.
10 John Aubrey, *Miscellanies upon Various Subjects* (London, 1857), p. 131.
11 Webster, *The Encyclopedia of Superstitions*, p. 165.
12 Sidney Oldall Addy, *Household Tales with Other Traditional Remains, Collected in the Counties of York, Lincoln, Derby, and Nottingham* (London and Sheffield, 1895), p. 101.
13 R. A. Leslie Moore, 'Indian Superstitions', *Journal of the Royal Society of Arts*, LIX/3040 (1911), p. 368.
14 See www.dnaindia.com, accessed 14 February 2019.
15 James C. Baker, 'Echoes of Tommy Knockers in Bohemia, Oregon, Mines', *Western Folklore*, XXX/2 (1971), p. 119.
16 See 'Mining Folklore', www.miningweekly.com, accessed 18 March 2019.
17 Addy, *Household Tales with Other Traditional Remains*, p. 141.
18 Daniel Lindsey Thomas and Lucy Blaney Thomas, *Kentucky Superstitions* (Princeton, NJ, 1920), p. 234.
19 Addy, *Household Tales with Other Traditional Remains*, p. 62.

Afterword

1 John Fowles, *The French Lieutenant's Woman* (London, 1969), p. 335.
2 Ibid., p. 327.
3 See Phoebe Weston, 'Britain Goes Week without Coal Power for First Time since Industrial Revolution', www.independent.co.uk, 8 May 2019.

SELECT BIBLIOGRAPHY

Andrews, Thomas G., *Killing for Coal: America's Deadliest Labor War* (Cambridge, MA, 2008)
Cassidy, Samuel M., ed., *Elements of Practical Coal Mining* (New York, 1973)
Church, Roy, with Alan Hall and John Kanefsky, *The History of the British Coal Industry*, vol. III: *1830–1913: Victorian Pre-eminence* (Oxford, 1986)
Davies, Alan, *The Pit Brow Women of Wigan Coalfield* (Stroud, 2006)
Francis, Hywel, and David Smith, *The Fed: A History of the South Wales Miners in the Twentieth Century* (London, 1980)
Freese, Barbara, *Coal: A Human History*, 2nd edn (New York, 2016)
Galloway, Robert, *Annals of Coal Mining and the Coal Trade*, vol. I (London, 1898)
Gray, Douglas, *Coal: British Mining in Art, 1680–1980*, exh. cat., organized by the Arts Council of Great Britain with the National Coal Board, City Museum and Art Gallery, Stoke-on-Trent, Glynn Vivian Art Gallery, Swansea, Science Museum London, D.L.I. Museum and Arts Centre, Durham, and Castle Museum, Nottingham (London, 1982)
Gray, Stephen, *Steam Power and Sea Power: Coal, the Royal Navy, and the British Empire, c. 1870–1914* (London, 2018)
Hart, Peggy M., *The Magic of Coal* (West Drayton, 1945)
Hatcher, John, *The History of the British Coal Industry*, vol. I: *Before 1700: Towards the Age of Coal* (Oxford, 1993)
J.C., *The Compleat Collier; or, The Whole Art of Sinking, Getting, and Working, Coal-mines, &c. As is now used in the Northern Parts, Especially about Sunderland and New-castle* (London, 1708)
Jevons, H. Stanley, *The British Coal Trade* (London, 1915)
John, Angela V., *By the Sweat of Their Brows: Women Workers at Victorian Coal Mines* (London, 2006)
Jones, Christopher F., *Routes of Power: Energy and Modern America* (Cambridge, MA, 2014)

Lawrence, D. H., *Sons and Lovers*, ed. David Trotter (Oxford, 2009)

Llewellyn, Richard, *How Green Was My Valley* (London, 1939)

Lloyd, A. L., ed., *Come All Ye Bold Miners: Ballads and Songs of the Coalfields* (London, 1952)

Long, Priscilla, *Where the Sun Never Shines: A History of America's Bloody Coal Industry* (New York, 1989)

McIvor, Arthur, and Ronald Johnston, *Miners' Lung: A History of Dust Disease in British Coal Mining* (Aldershot, 2007)

McManners, Robert, and Gillian Wales, *Shafts of Light: Mining Art in the Great Northern Coalfield* (Durham, 2002)

Nef, J. U., *The Rise of the British Coal Industry*, 2 vols (London, 1932)

Orwell, George, *The Road to Wigan Pier* (London, 1937)

Sinclair, Upton, *King Coal* (New York, 1917)

Smith, Martin Cruz, *Rose* (London, 2014)

Thesing, William B., ed., *Caverns of Night: Coal Mines in Art, Literature, and Film* (Columbia, SC, 2000)

Verne, Jules, *The Child of the Cavern; or, Strange Doings Underground* (London, n.d.)

Watt, Alexander, *The History of a Lump of Coal: From the Pit's Mouth to a Bonnet Ribbon* (London, 1882)

Wells, Harold C., *The Earth Cries Out* (Sydney, 1950)

Zola, Émile, *Germinal*, trans. Peter Collier, intro. Robert Lethbridge (Oxford, 2008)

ASSOCIATIONS AND WEBSITES

Big Pit National Coal Museum, Wales
www.museum.wales/bigpit
Former coal mine turned award-winning national museum in Blaenafon.

European Route of Industrial Heritage
www.crih.nct
A gigantic network of the most important industrial sites – including many coal-industry sites – across Europe. The website includes anchor points (key sites), regional routes and theme routes. The mining theme route features numerous coal mines.

Healey Hero
www.healeyhero.co.uk
The website is named for Philip Healey, a member of Ilkeston Mines Rescue, 1954–71. The idiosyncratic site includes a comprehensive glossary of pit terminology and words specific to the coal-mining industry.

International Mines Rescue Body
www.minerescue.org
An informal association representing mine rescue organizations from around the world.

Kentucky Coal Education
www.coaleducation.org
Provides information about the coal industry for students, teachers and the general public.

Kentucky Geological Survey
www.uky.edu/KGS/coal
This section of the Kentucky Geological Survey website contains a wealth of scientific information about coal.

National Coal Mining Museum for England
www.ncm.org.uk
This museum is located at the site of Caphouse Colliery, near Wakefield,
Yorkshire.

National Mining Museum Scotland
www.nationalminingmuseum.com
Scotland's national coal-mining museum, located at the site of the Lady
Victoria Colliery, Newtongrange.

Network of European Coal Mining Museums
www.europeancoalminingmuseums.com
An independent voluntary association with one coal-mining museum
per country, pursuing its aims by educational and research methods.

Northern Mine Research Society
www.nmrs.org.uk
The group maintains a website dedicated to the preservation and
recording of mining history, and publishes books under the series
title British Mining.

Show Caves
www.showcaves.com
A comprehensive index to caves, karst features, springs, mines, gorges
and other subterranea around the world, maintained by Jochen Duckeck.
Includes a detailed subsection on show mines, including coal mines.

Underground Coal
www.undergroundcoal.com.au
A website covering all aspects of underground coal mining in Australia.

World Coal
www.worldcoal.com
The website carries coal news from around the world.

World Coal Association
www.worldcoal.org
The World Coal Association is the global network for the coal industry.
Formed of major international coal producers, it works to demonstrate
and gain acceptance for the fundamental role coal plays in achieving a
sustainable, low-carbon future.

ACKNOWLEDGEMENTS

This book was supported by the University of Tasmania, which provided funding for various research trips to coal mines and other coal tourism sites in Australia, Britain, Europe and the U.S., and granted me a period of study leave during which the bulk of the writing was completed. I am very grateful for that support.

I would like to thank friends and colleagues who read drafts of all or part of this book for their detailed feedback: in particular, Elizabeth Leane (author of *South Pole*), who read and offered sagacious advice on early drafts of several chapters as well as the completed manuscript, and Lisa Fletcher (with whom I co-authored *Cave*) for her astute comments on the literature chapter. I am grateful to Matthew Stillwell, who gave me the Chinese poster (reproduced on p. 6) from the collection of Professor Bruce Johnson, and to my colleague Isabel Wang, who translated the text it carries for me. I am also grateful to Rick Snell, who gave me a copy of Harold C. Wells's *The Earth Cries Out*. I would especially like to thank Dan Savage (Artstop Studios) for generously allowing me to reproduce an image of one of his artworks; the Yamamoto family for permission to use an image of a work by Sakubei Yamamoto; and Andy Irvine for kindly allowing me to reproduce lyrics from one of his songs. Further, I would like to acknowledge the assistance of Rachel Adams and Susanna Hennighausen in the Document Delivery section of the University of Tasmania's Morris Miller Library, who sourced a number of illustrations for me; the assistance of Curator Emeritus John Liffen, who shared his knowledge of the Mining Gallery at the Science Museum, London; the expertise of Richard Williams, who took the photographs of some of my coal memorabilia; and the company of Joy Crane, Radhika Mohanram and Kathleen Crane, who (with varying degrees of enthusiasm) visited coal mines with me.

It has been a pleasure to work with the people at Reaktion Books again. I am grateful in particular to Michael Leaman and Daniel Allen for the opportunity to write a second book in the wonderful Earth series, and to Amy Salter and Susannah Jayes for their great work during the publication process. I'd also like to thank Peter A. Shulman, Associate Professor of History at the Case Western Reserve University, for his detailed reader's report.

My wife Joy, as ever, has been extremely supportive during the research for and writing of this book. Along the way she has learnt far more about coal than I suspect she ever wanted to know; she still doesn't understand why I chose coal as a topic.

Permission

The four lines from Andy Irvine's song 'Hard Times in 'Comer's Mines' (Andy Irvine and Luke Plumb, *Precious Heroes*, 2016) on p. 130 are reproduced by kind permission of Andy Irvine.

PHOTO ACKNOWLEDGEMENTS

The author and the publishers wish to express their thanks to the below sources of illustrative material and /or permission to reproduce it.

AAP Photos: p. 68 (Mick Tsikas); author's collection: pp. 6, 12, 15, 34, 38, 55 (Richard Williams), 72 (Richard Williams), 77, 90, 91 (Richard Williams), 105 (Richard Williams), 113, 115, 117 bottom left, 130 (reproduced by kind permission of Andy Irvine), 131, 164 (Richard Williams), 165, 168; The British Library, London: pp. 18, 88, 126; © The Trustees of the British Museum: pp. 144, 145; Ralph Crane: pp. 10, 13, 21, 25, 53 centre, 60, 61, 62, 64–5, 66, 82, 85, 86, 93, 94, 95, 97, 98, 102, 103, 134, 142, 160, 161, 170, 173; Dutch National Archives: p. 41 (Harry Pot); Courtesy of Stephen Greb, Kentucky Geological Survey, University of Kentucky: p. 14; Kröller-Müller Museum, Otterlo: p. 148 bottom; Lancashire Mining Museum at Astley Green: p. 152; Library of Congress, Washington, DC: pp. 35, 167; Marcher Museum, Berlin: p. 159; The Metropolitan Museum of Art, New York: pp. 9, 27; Meunier Museum, Brussels: pp. 135, 149; Musée d'Orsay, Paris: p. 147; The National Archives, London: p. 151 (INF3/156); National Archives and Records Administration, Washington, DC: pp. 51, 78, 137, 155; photograph courtesy of the National Coal Mining Museum for England: p. 121; National Gallery of Art, Washington, DC: p. 146 bottom; National Library of Australia: pp. 46, 59; National Library of France: p. 117 bottom right; Nottingham Playhouse: p. 124 (Darren Bell); Old Masters Picture Gallery, Dresden: p. 24; Collection de la Province de Hainaut, Charleroi: p. 154 (BPS22); Recap Data/World Energy Council: p. 39; image courtesy of the Royal Exchange Theatre, Manchester: p. 125; image courtesy of Dan Savage, Artstop Studios: p. 74; © The Board of Trustees of the Science Museum: p. 139; Alexander Turnbull Library, Wellington: p. 40 (Albert Percy Godber); Victoria and Albert Museum, London: p. 44; Wellcome Collection, London: p. 30; Yale Center for British Art, New Haven: p. 146 top; © Yamamoto Family, courtesy of Tagawa City Coal Mining Historical Museum: p. 158.

INDEX

Page numbers in *italics* refer to illustrations